MW00462272

MANAGING WORKERS' COMPENSATION

A Guide to Injury Reduction and Effective Claim Management

Occupational Safety and Health Guide Series

Series Editor

Thomas D. Schneid
Eastern Kentucky University
Richmond, Kentucky

Published Titles

Creative Safety Solutions
by Thomas D. Schneid

Occupational Health Guide to Violence in the Workplace
by Thomas D. Schneid

Motor Carrier Safety: A Guide to Regulatory Compliance
by E. Scott Dunlap

Disaster Management and Preparedness
by Thomas D. Schneid and Larry R. Collins

**Managing Workers' Compensation: A Guide to Injury Reduction
and Effective Claim Management**
by Keith R. Wertz and James J. Bryant

Forthcoming Titles

Physical Hazards of the Workplace
by Larry R. Collins

MANAGING WORKERS' COMPENSATION
A Guide to Injury Reduction and Effective Claim Management

Keith R. Wertz
James J. Bryant

LEWIS PUBLISHERS
Boca Raton London New York Washington, D.C.

Library of Congress Cataloging-in-Publication Data

Wertz, Keith R.
  Managing workers' compensation : a guide to injury reduction and effective claim
management / Keith R. Wertz and James J. Bryant.
    p.   cm.-- (Occupational safety and health guide series)
  Includes index.
  ISBN 1-56670-348-4
    1. Workers' compensation--United States. 2. Industrial safety--United States. 3.
Workers' compensation claims--United States. I. Bryant, James J. II. Title. III. Series.

HD7103.65.U6 W46 2000
658.3'254—dc211
                                                                              00-046348
                                                                              CIP

© 2001 by CRC Press LLC
Lewis Publishers is an imprint of CRC Press LLC

No claim to original U.S. Government works
International Standard Book Number 1-56670-348-4
Library of Congress Card Number 00-046348
Printed in the United States of America   4 5 6 7 8 9 0
Printed on acid-free paper

About the Authors

Keith R. Wertz

Mr. Wertz earned an M.S. in Loss Prevention and Safety from Eastern Kentucky University. He is an active member of the Bluegrass Chapter of the American Society of Safety Engineers and the Kentucky Safety and Health Network. Additionally, Mr. Wertz serves on the Expert Advisory Panel of Safety On-Line.

As the Loss Control Manager for Midwestern Insurance Alliance, where he has been employed since October 1995, Mr. Wertz oversees loss control services provided to insured companies throughout the U.S. During his tenure with Midwestern Insurance Alliance, Mr. Wertz has conducted in excess of 1000 consultative visits with insured companies, instructing business owners, managers, and supervisors how to minimize workers' compensation costs through injury prevention techniques and cost containment strategies. Prior occupational safety experience includes the appointment as District Safety Coordinator with Borg Warner Corporation and performing contract loss control services for several multi-line insurance carriers.

Through his experience providing consultative loss control services to a wide variety of industries, Mr. Wertz has gained expertise in the identification of internal management practices that fail to produce injury prevention and cost containment results, and is adept at communicating sound loss control principles to individuals who are responsible for workers' compensation cost control. His straightforward and unambiguous approach to addressing the relationship between occupational safety, cost containment, and workers' compensation insurance is embodied in this text.

James J. Bryant

Mr. Bryant earned an M.S. in Safety Management from Murray State University and an M.Ed. in Occupational Education from the University of Louisville. A member of the Purchase Area Chapter of the American Society of Safety Engineers, he is a Certified Hazardous Materials Manager and has obtained designation as an Associate Safety Professional.

Employed with Midwestern Insurance Alliance since February 1999 as a Loss Control Representative, his previous safety experience includes a 20-year tenure as a commissioned officer in the U.S. Army, incorporating safety management into inherently dangerous operations and as a Safety Technician at Murray State University. He also serves as an adjunct instructor in the Murray State University Occupational Safety and Health Training Center and the Department of Occupational Safety and Health.

The combination of Mr. Bryant's formal education and diverse career endeavors has afforded him a unique opportunity to observe the necessity and success of systemic applications of safety management for controlling loss. This insight is described in this text. Mr. Bryant has also authored articles pertinent to this subject, which have been published in *Professional Safety* and *The Synergist*.

Contents

Appendices

chapter one

Workers' compensation overview

Contents

1.1   Introduction

Workers' compensation — the words alone are enough to elicit a reaction, and too often it is a negative one. From the outset, that negative connotation challenges all who are involved with the system. Consider, for example, the following statements and compare them to your own involvement with workers' compensation insurance: It is unpleasant and inconvenient, a business expense and not a profit maker. Many cases are antagonistic — either between the injured employee and the employer, or between the injured employee and the insurance provider, or between the employer and the insurance provider, or between all parties. An injury has occurred and somebody is literally in pain. The company's safety record and insurance rating will be tarnished. Processing the claim only means more work for the individual charged with this responsibility. Factor in the rising costs of medical services and it's easy to see how workers' compensation becomes an object of discontent.

As is the case with so many other legislative efforts and programs, which are initiated with noble intent, it's not supposed to be this way. Overall, the purpose of workers' compensation insurance has been to minimize the financial impact on the employee and the employer that may be incurred as the result of occupational injuries. Some will argue that since its inception, workers' compensation legislation has been politically swayed by labor, by management, and even by the insurance industry. The current situation appears to be one in which no one is satisfied. Those issues are not addressed in this text. Instead, proven methods of "making the best" of the current situation are discussed.

1.2   Basic principles

The workers' compensation system revolves around the concept that the employer agrees to pay for all injuries to employees regardless of fault, and, in turn, the employee agrees to give up his or her right to sue the employer in relation to an on-the-job injury. This concept is based on the opinion that society has a moral responsibility to care for workers injured at work and their families.

One of the things that make it difficult to generalize about workers' compensation is that it is a state-based system. For almost any possible statement about workers' compensation, one can find a state that does things differently. As a result, most generalizations about workers' compensation have exceptions. However, the key elements common to almost all states are

- All work-related injuries and illnesses must be compensated, regardless of fault.
- Compensation is limited to
  - medical costs,
  - indemnity payments, which can take the form of temporary disability payments or permanent disability payments,
  - rehabilitation, and
  - defined death benefits.
- One of the major limitations of workers' compensation is that the system does not allow payment for pain and suffering nor may workers receive punitive damages from their employers.
- Since it is a state-based system, rules can vary among states.

1.3   Historical perspective

Understanding the workers'compensation system requires an appreciation of the history of occupational safety, because workers' compensation is an outgrowth of that experience. What we have today is a result of what occurred in the past. Realizing that specific events in the historical industrial development were the causal factors for today's compensation system provides insight into the situation, although it may not clarify all issues.

Workers' compensation insurance in the U.S. originated as a turn-of-the-century reform of the insurance system designed to maximize benefits to workers while minimizing administrative and litigation costs. The essence of the workers' compensation system is that the employer must compensate the worker for all work-related injuries and illnesses regardless of fault. The compensation the worker can receive is limited by statute, and the worker cannot sue in court for further damages.

Occupational safety did not begin with the passage of the Occupational Safety and Health Act of 1970 (OSHA). Several private and public actions have led to the current approach to occupational safety and the compensation of injured workers. Listed below are some significant events in occupational safety in the U.S. Notice the intermittent but persistent efforts to address workers' compensation.

1812 — the embargo created by the War of 1812 spurred the development of the New England textile industry and the founding of factory mutual companies. These early insurance companies inspected properties for hazards and suggested loss control and prevention methods in order to secure low rates for their policyholders.

1864 — the Pennsylvania Mine Safety Act was passed into law.

1864 — North America's first accident insurance policy was issued.

1867 — Massachusetts instituted the first government-sponsored factory inspection program.

1877 — Massachusetts passed a law requiring guarding for dangerous machinery and took authority for enforcement of factory inspection programs.

1878 — the first recorded call by a labor organization for federal occupational safety and health law was heard.

1896 — an association to prevent fires and write codes and standards, the National Fire Protection Association, was founded.

1902 — Maryland passed the first workers' compensation law.

1904 — the first attempt by a state government to force employers to compensate their employees for on-the-job injuries was overturned when the Supreme Court declared Maryland's workers' compensation law unconstitutional.

1911 — a professional technical organization responsible for developing safety codes for boilers and elevators, the American Society of Mechanical Engineers, was founded.

1911–1915 — 30 states passed workers' compensation laws.

1911 — the American Society of Safety Engineers, dedicated to the development of accident prevention techniques and the advancement of safety engineering as a profession, was founded.

1912 — a group of engineers representing insurance companies, industry, and government met in Milwaukee to exchange data on accident prevention; the organization formed at this meeting became the National Safety Council.

1916 — the Supreme Court upheld the constitutionality of state workers' compensation laws.

1918 — the American Standards Association, now known as the American National Standards Institute, was founded. It has been responsible for the development of many voluntary safety standards, some of which are referenced in current laws.

1936 — Frances Perkins, Secretary of Labor, called for a federal occupational safety and health law, an action that came a full 58 years after organized labor's first recorded request for a law of this nature.

1936 — the Walsh-Healey Act was passed, which required that all federal contracts be fulfilled in a healthful and safe working environment.

1948 — all 48 states had workers' compensation laws.

1952 — the Coal Mine Act was passed into law.

1960 — specific safety standards were promulgated for the Walsh-Healey Act.

1966 — the Metal and Nonmetallic Mines Safety Act was passed.

1966 — the U.S. Department of Transportation and its sections, the National Highway Traffic Safety Administration and the National Transportation Safety Board, were established.

1968 — President Johnson called for a federal occupational safety and health law.

1969 — the Construction Safety Act was passed.

1969 — the Board of Certified Safety Professionals, which certifies practitioners in the safety profession, was established.

1970 — President Nixon signed into law the Occupational Safety and Health Act, thus creating the Occupational Safety and Health Administration and the National Institute for Occupational Safety and Health.

1.4   Methods of coverage

Workers' compensation insurance can be provided in several ways. The method of coverage is determined to some degree by state laws, and also by the availability of insurance providers, the financial solvency of the company, the particular business activity, and the loss experience of the company. Examples of coverage sources, or categories, include:

Traditional commercial insurance companies — commercial insurance companies that market workers' compensation insurance. Basically, there are two classifications: voluntary and involuntary programs. In voluntary programs, policies are sold to businesses with an adequate premium and safety program to make the policy worthwhile to the insurance company. Involuntary programs involve small employers and/or employers with a high injury rate. This grouping is known as an assigned risk pool because these employers are assigned to an insurance company, which usually provides insurance coverage at a higher rate commensurate with the higher risk.

Self-insurance funds — employers, usually in a common business (i.e., lumber and building materials dealers), who pledge their individual assets to collectively guard against exposure for paying workers' compensation benefits. These funds have become increasingly popular because of their availability and potential for saving premium costs by controlling workers' compensation costs.

Individual self-insured — large employers may have the option to be self-insured, which allows for employers to provide workers' compensation coverage by paying the benefits themselves.

1.5 Key participants

Several individuals and organizations are involved in the workers' compensation matrix. As displayed in Figure 1.1, this relationship includes many more participants than just the employee, the employer, and the workers' compensation insurance company.

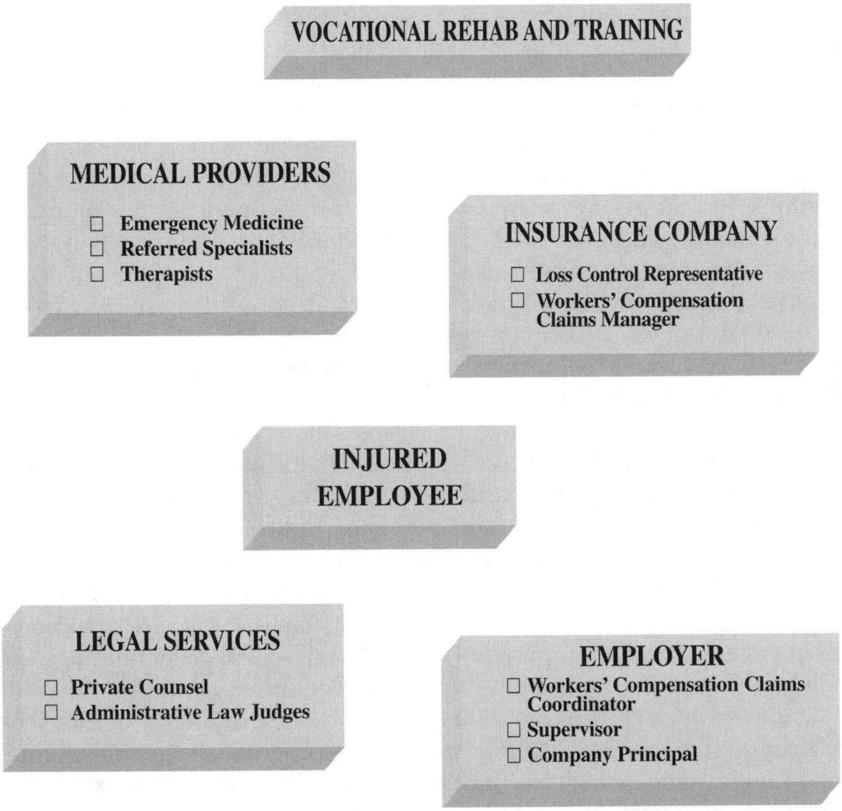

VOCATIONAL REHAB AND TRAINING

MEDICAL PROVIDERS
- Emergency Medicine
- Referred Specialists
- Therapists

INSURANCE COMPANY
- Loss Control Representative
- Workers' Compensation Claims Manager

INJURED EMPLOYEE

LEGAL SERVICES
- Private Counsel
- Administrative Law Judges

EMPLOYER
- Workers' Compensation Claims Coordinator
- Supervisor
- Company Principal

Figure 1.1

The intricacy of the injury and subsequent developments dictate who will become involved. Regardless of how many do become involved, communications among these participants is critical to the timely and efficient resolution of claims. The broader the involvement of participants, the more challenging it becomes to establish and maintain effective communications. Communication is so important that it will surface throughout this text.

To increase this challenge, routine turnover of employees, change in workers' compensation insurance providers (through competitive premiums, service, etc.), changes in the medical industry, associated changes in administrative processes, and changes in workers' compensation laws all contribute to keeping the process dynamic. Just when you think you have it all figured out and under control, along comes another change.

1.6   Facts vs. fiction

Quite often an individual's understanding of the workers' compensation system is flawed, based upon fictitious information or opinions rather than fact. This situation complicates the management of workers' compensation because the perceptions must be addressed; otherwise, they will undermine the program. Review the myths below and see if they are held as reality in your organization.

The insurance company or the state pays for all expenses associated with workers' compensation. If this were the case, why would you pay a monthly premium to that insurance company? Insurance companies are motivated by the same principle of business that motivates all other companies — profit. They assess premiums based upon the perception of risk involved with a potential insured. The insurance company does not want to spend more on claims than they can recoup on premiums. So, yes, the insurance company does pay for workers' compensation expenses, but with the money that has been collected in premiums.

Employees do not pay for workers' compensation. This statement is true if you are referring only to direct costs. However, it's important to understand the difference between direct and indirect costs. Workers' compensation is a business expense, no different than the rent, utility, or tax bills. Employees do not directly pay for these expenses; however, they do pay for them indirectly. If the business expenses are not controlled, then the profits that would have been available for reinvestment (facility, salaries, equipment) are not there. Employees may not receive bonuses or profit sharing. Ultimately, employees do pay for workers' compensation insurance.

All workers' compensation claims are false. This perception is one of those that can undermine your organization's safety program and even the morale of the entire work force. It's endemic in organizations where feelings of distrust are manifested in other areas. Usually these organizations are experiencing high employee turnover, absenteeism, low morale, no loyalty to the company, etc. The fact is that some people do get hurt at work. To imply that all claims are false is prejudicial and unfounded. The supervisor and employer should investigate all accidents to determine the cause and to preclude its recurrence. In doing so, the validity of the claim will usually be established. Most often, when supervisors and management feel that all workers' compensation claims are false it is because their opinions are based upon assumptions rather than data obtained through accident investigations.

Accidents are the fault of the employee. Some, but not all, accidents are the fault of the employee but the cause cannot be determined unless the accident is investigated. And what if the accident is the employee's fault? Is it not compensable? Certain provisions in the definition of workers' compensation insurance do exclude some injuries from coverage. (However, it is not the specific injury that is not covered; rather, the circumstances surrounding the injury cause it to be excluded.) One of the basic principles of workers' compensation is all work-related injuries and illnesses must be compensated, regardless of fault. There are occasions when accidents and injuries result from conditions in the workplace, equipment in the workplace, or the managerial approach to operations.

The more employees know about workers' compensation, the more they will abuse it. Again, this mind-set is indicative of distrust. The employer does not trust the employees with information. As already stated, when this relationship exists in an organization, there will be other symptoms of distrust — high employee turnover, low morale, absenteeism, poor quality control, etc. This same argument is commonly offered as a reason not to educate or train employees regarding ergonomic hazards. It's an unfounded fear on behalf of the employer, and it's an insult to the employee. The implication is that employees are motivated to abuse the system. Granted, some will attempt to do so, but the establishment and management of an inclusive, formal safety program can preclude this from occurring. (Information regarding safety program management is included in Chapter 2.)

An accident must be witnessed to be compensable. This is not true. What is true is that as more information is gathered during the accident investigation, the greater is the confidence that the events surrounding the accident were determined, and, therefore, its compensability is better determined. Witnesses enhance this process, as long as they are honest. However, if an accident is not witnessed, it is not automatically denied.

Reporting an injury to the insurance carrier is an endorsement and validation of the facts, as recorded on the first-report-of-injury form, by the insured. This is another untruth. The first report-of-injury form is the document that begins the administrative audit trail for the workers' compensation claim. It has a short suspense (usually must be submitted within 24 hours of the injury). Therefore, it is understood that the information included may not be thorough. Subsequent accident investigation will ferret out as much information as possible and will be incorporated into the compensability decision-making process. Submission of the first-report-of-injury form is an initiating event — it is not a concluding event.

If it happened at work, it must be compensable. This statement presents an absolute, and there are few, if any, absolutes in workers' compensation. The occurrence of an injury at work is a very important compensability determinant. But there may be other circumstances to be considered, such as whether or not the employee was under the influence of drugs or alcohol, or if the employee was in direct and willful violation of company policy and procedures. Workers' compensation compensability is not a "blank check" with total coverage of all injuries that happen at work.

A preexisting condition will disqualify a claimant from receiving benefits. This statement needs clarification. If an employee continues to suffer from an injury that was covered by another workers' compensation claim (and/or carrier), then there exists the possibility that the claim will be subrogated to the previous carrier. However, the mere preexistence of a condition that is aggravated at work does not exclude the claimant from receiving benefits.

These examples demonstrate the confusion that can arise when attempting to make "one-size-fits-all" statements about workers' compensation. Add to this confusion the fact that money is involved, and the situation is ripe for myths to develop and flourish.

1.7   What are the costs?

All firms suffer from the high cost of workers' compensation and are better off if the injury and illness rates, as a whole, decline. There are also indirect and hidden costs associated with injuries and illnesses. In the past decade, the average cost of a serious workers' compensation injury has risen over 300% for lost wages and over 400% for medical costs. As a nation, businesses are paying billions of dollars for workers' compensation–related expenses. How much does it cost your business?

Clearly, the costs of workers' compensation have risen steadily through the years — both the number of claims and the average costs per claim have increased. Moreover, the costs of both the medical and disability components of workers' compensation have increased at rates far greater than inflation. These rapid rates of increase have caused many states to focus on ways of reforming the system, including ways to provide greater incentives for employers to reduce accidents and injuries.

A few specific cost drivers affect workers' compensation, all of which are addressed in other sections of this book, with recommended methods of countering their upward trend. But it's important to understand early on their contribution to the rising costs of workers' compensation and also to understand that they have a synergistic relationship.

- Increased wages — As wages increase, workers' compensation premium also increases. If wages increase 10%, the premium will increase proportionately. When employees receive lost-wage benefits, they are compensated a percentage of their salary (determined by individual state laws). Therefore, as salaries increase, lost-wage benefits increase.
- Longer time off the job — Typically, lost-time benefits are not paid unless an employee misses an established number of days of work

due to the injury. Unfortunately, this appeals to some employees as an opportunity to remain off work the appropriate amount of time to qualify for lost-time benefits. Obviously, this involves an individual's integrity, but it also involves the company's philosophy and approach to early-return-to-work policies and procedures.

- Medical costs — These costs have risen dramatically, and, in some cases, increased diagnostic tests and treatments have been rendered. Medical costs now exceed 40% of all benefits paid (for work-related injuries, all medical costs are paid).
- Litigation — Because of the dollars involved and the ambiguities in many areas of the workers' compensation laws and regulations, many claims are contested by employees, employers, and insurance companies. The cost of these legal proceedings and the fees paid to attorneys add to the cumulative expense of workers' compensation.
- Administrative costs — As premiums increase, as workers' compensation payments increase, and as litigation increases, administrative costs increase. Also, as premiums increase there exists the probability that the profit realized by the insurance provider will increase.
- Fraud — Dishonest individuals may fake an injury or describe its symptoms as more serious than they really are. Medical providers may over-treat an injury or illness. Finally, attorneys may complicate a scenario. Often insurance providers will contract the services of investigation agencies to counter fraudulent claims. These services cost, as well.

1.8 Summary

Workers' compensation represents a significant business expense for employers, in some cases 20–30% of their payroll. The system, however, is critical to the economic well-being of the injured worker. There is a clear trend of increased benefits to the injured and increased cost to the employer, costs which are ultimately passed on to consumers in the form of higher prices.

To a large extent, employers can influence their net costs for this coverage, positively or negatively, through their own efforts. The system contains sufficient cost incentives to motivate management to contain costs. However, the enormity of the cost and the means of controlling them must be understood, accepted, and applied. When health and safety are high priorities with management, then employers, employees, and the public realize direct economic benefits.

chapter two

Avoiding claims through safety management

Contents

2.1   Manage or be managed

Quantification of the cost of only one back injury that requires surgical intervention compared to the administrative overhead of managing a safety program can dramatically demonstrate why safety management is important. This simple cost-benefit analysis is even more conclusive when one considers the availability of safety management programs and the services of loss control professionals who will assist in the establishment and maintenance of them. If an organization does not manage its occupational safety and health program, inclusive of injuries and workers' compensation, then that organization will be managed by those losses.

The following paragraphs are an excerpt from OSHA's Unified Agenda for Long Term Actions as it appeared in the *Federal Register*, April 24, 2000. This portion of the Agenda was discussing OSHA's philosophy and intentions regarding the ultimate proposal for a standard that regulates the creation and utilization of formal safety and health programs, specifically the statement of need and the risks involved. Understand that OSHA has nearly three decades of experience from which to draw this conclusion. Developed from visiting business entities in all sectors of the economy and in all regions of the country, their perspective can be described as a true sampling of the beneficial effects of a safety and health program, or the negative effects of its absence.

Worksite-specific safety and health programs are increasingly being recognized as the most effective way of reducing job-related accidents, injuries, and illnesses. Many states have to date passed legislation and/or regulations mandating such programs for some or all employers, and insurance companies have also been encouraging their client companies to implement these programs, because the results they have achieved have been dramatic. In addition, all of the companies in OSHA's Voluntary Protection Programs (VPP) have established such programs and are reporting injury and illness rates that are sometimes only 20% of the average for other establishments in their industry. Safety and health programs apparently achieve these results by actively engaging front-line employees, who are closest to operations in the workplace and have the highest stake in preventing job-related accidents, in the process of identifying and correcting occupational hazards. Finding and fixing workplace hazards is a cost-effective process, both in terms of the avoidance of pain and suffering and the prevention of the expenditure of large sums of money to pay for the direct and indirect costs of these injuries and illnesses. For example, many employers report that these programs return between $5 and $9 for every dollar invested in the program, and almost all employers with such programs experience substantial reductions in their workers' compensation premiums. OSHA believes that having employers evaluate the job-related safety and health hazards in their workplace and address any hazards identified before they cause occupational injuries, illnesses, or deaths is an excellent example of "regulating smarter," because all parties will benefit: workers will avoid the injuries and illnesses they are currently experiencing; employers will save substantial sums of money and increase their productivity and competitiveness; and OSHA's scarce resources will be leveraged as employers and employees join together to identify, correct, and prevent job-related safety and health hazards.

Workers in all major industry sectors in the United States continue to experience an unacceptably high rate of occupational fatalities, injuries, and illnesses. For 1996, the Bureau of Labor Statistics reported that 6.2 million injuries and illnesses occurred within private industry. For 1997, BLS reported that 6,218 workers lost their lives on the job. There is increasing evidence that addressing hazards in a piecemeal fashion, as employers tend to do in the absence of a comprehensive safety and health program, is considerably less effective in reducing accidents than a systematic approach. Dramatic evidence of the seriousness of this problem can be found in the staggering workers' compensation bill paid by America's employers and employees: about $54 billion annually. These risks can be reduced by the implementation of safety and health programs, as evidenced by the experience of OSHA's VPP participants, who regularly achieve injury and illness rates averaging one-fifth to one-third those of competing firms in their industries. Because the proposed rule addresses significant job-related hazards, the rule will be effective in ensuring a systematic approach to the control of long-recognized hazards, such as lead, which are covered by existing OSHA standards, and emerging hazards, such as lasers and violence in the workplace, where conditions in the workplace would require control under the General Duty Clause of the Act.

2.2   OSHA safety and health program

In 1989, OSHA issued recommended guidelines for the effective management and protection of worker safety and health, and the above-referenced draft standard was modeled on these guidelines. The original guidelines, as contained in OSHA's *General Industry Digest*, are described below. To add credence to these OSHA recommendations, it has been the experience of the authors during hundreds of loss control consultative visits that, time and time again, those organizations that incorporated some or all of these guidelines enjoyed a more productive and less injurious work environment.

2.2.1   General

Employers are encouraged to institute and maintain a program that provides adequate systematic policies, procedures, and practices that protect their employees from, and allow them to recognize, job-related safety and health hazards. An effective program includes provisions for the systematic identification, evaluation, and prevention or control of general workplace hazards, specific job hazards, and potential hazards that may arise from foreseeable conditions. Although compliance with the law, including specific OSHA standards, is an important objective, an effective program looks beyond specific requirements of law to address all hazards. It will seek to prevent injuries and illnesses, whether or not compliance is at issue. The extent to which the program is described in writing is less important than how effective it is in practice. As the size of a workplace or the complexity of a hazardous operation increases, however, the need for written guidance increases to ensure clear communication of policies and priorities as well as a consistent and fair application of rules.

2.2.2   Major elements

An effective occupational safety and health program includes four main elements:

- Management commitment and employee involvement
- Worksite analysis
- Hazard prevention and control
- Safety and health training

Each of these elements is discussed below.

2.2.2.1 Management commitment and employee involvement

Management commitment and employee involvement are complementary and form the core of any occupational safety and health program. Management's commitment provides the motivating force and the resources for organizing and controlling activities within an organization. In an effective program, management regards worker safety and health as a fundamental value of the organization and applies its commitment to safety and health protection with as much vigor as to other organizational goals. Employee involvement provides the means by which workers develop and/or express their own commitment to safety and health protection for themselves and for their fellow workers.

In implementing a safety and health program, there are various ways to provide commitment and support by management and employees. Some recommended actions are described briefly below:

- State clearly a worksite policy on safe and healthful work and working conditions, so all personnel with responsibility at the site (and personnel at other locations with responsibility for the site) fully understand the priority and importance of safety and health protection in the organization.
- Establish and communicate a clear goal for the safety and health program and define objectives for meeting that goal, so all members of the organization understand the results desired and the measures planned for achieving them.
- Provide visible top management involvement in implementing the program, so all employees understand that management's commitment is serious.
- Arrange for and encourage employee involvement in the structure and operation of the program and in decisions that affect their safety and health, so they will commit their insight and energy to achieving the safety and health program's goals and objectives.
- Assign and communicate responsibility for all aspects of the program, so managers, supervisors, and employees in all parts of the organization know what performance is expected of them.
- Provide adequate authority and resources to responsible parties, so assigned responsibilities can be met.
- Hold managers, supervisors, and employees accountable for meeting their responsibilities, so essential tasks will be performed.
- Review program operations at least annually to evaluate success in meeting goals and objectives, so deficiencies can be identified and the program and/or the objectives can be revised if they do not meet the goal of effective safety and health protection.

2.2.2.2 Worksite analysis

A practical analysis of the work environment involves a variety of worksite examinations in order to identify existing hazards and conditions and operations in which changes might occur to create new hazards. Unawareness of a hazard stemming from failure to examine the worksite is a sign that safety and health policies and/or practices are ineffective. Effective management actively analyzes the work and worksite to *anticipate* and thereby prevent harmful occurrences. In order that all hazards and potential hazards are identified, the following measures are recommended:

- Conduct comprehensive baseline worksite surveys for safety and health and periodic comprehensive update surveys.
- Analyze planned and new facilities, processes, materials, and equipment.
- Perform routine job hazard analyses.
- Conduct regular site safety and health inspections so new or previously missed hazards and failures in hazard control are identified.
- Provide a reliable system for employees to notify management personnel about conditions that appear hazardous and to receive timely and appropriate responses and encourage employees to use the system without fear of reprisal. This system utilizes employee insight and experience in safety and health protection and allows employee concerns to be addressed.
- Investigate accidents and "near-miss" incidents so their causes and the means for prevention can be identified.
- Analyze injury and illness trends over time so patterns can be identified and prevented.

2.2.2.3 Hazard prevention and control

Where feasible, workplace hazards are prevented by effective design of the job site or job. Where it is not feasible to eliminate such hazards, to prevent unsafe and unhealthful exposure they must be controlled in a timely manner once they are recognized. Specifically, as part of the program, employers should establish procedures to correct or control present or potential hazards in a timely manner. These procedures should include measures such as the following:

- Use engineering techniques where feasible and appropriate.
- Establish, at the earliest time, safe work practices and procedures that are understood and followed by all affected parties. Understanding and compliance are a result of training, positive reinforcement, correction of unsafe performance, and, if necessary, enforcement through a clearly communicated disciplinary system.
- Provide personal protective equipment (PPE) when engineering controls are not feasible.

- Use administrative controls, such as reducing the duration of exposure.
- Maintain the facility and equipment to prevent equipment breakdowns.
- Plan and prepare for emergencies, and conduct training and emergency drills, as needed, to ensure that proper responses to emergencies will be "second nature" for all persons involved.
- Establish a medical program that includes first aid onsite as well as nearby physician and emergency medical care to reduce the risk of any injury or illness that occurs.

2.2.2.4 Safety and health training

Training is an essential component of an effective safety and health program, addressing the safety and health responsibilities of both management and employees at the site, salaried and hourly. It is often most effective when incorporated into other education on performance requirements and job practices. The complexity of training depends on the size and complexity of the worksite as well as the characteristics of the hazards and potential hazards at the site.

2.2.2.4.1 *Employee training.* Employee training programs should be designed to ensure that all employees understand and are aware of the hazards to which they may be exposed and of the proper methods for avoiding such hazards.

2.2.2.4.2 *Supervisory training.* Supervisors should be trained to understand the key role they play in job site safety to enable them to carry out their safety and health responsibilities effectively. Training programs for supervisors should include the following:

- Analyzing the work under their supervision to anticipate and identify potential hazards.
- Maintaining physical protections in their work areas.
- Reinforcing employee training on the nature of potential hazards in their work and on needed protective measures, through continual performance feedback and, if necessary, through enforcement of safe work practices.
- Understanding their safety and health responsibilities.

2.2.3 Current modifications

In addition to the above elements, program evaluation has also been greatly emphasized. This evaluation is "double-edged" — not only should all hazard controls (engineering, administrative, or personal protective equipment) implemented to reduce risk be evaluated to determine their effectiveness, but also the safety program itself should be periodically evaluated for effectiveness. Signs and symptoms of gaps in a program can include

- Accidents or near-miss incidents that go unreported,
- Accidents or near-miss incidents that are not investigated,
- The absence of a preventive maintenance program for equipment and facilities,
- Poor, or nonexistent, internal communications in reference to safety issues,
- Upward trend in the organization's incident rate, and
- Employee/supervisor attitude that production supersedes safety.

2.2.4   Platform for success

Regardless of the format or the level of sophistication, organizations that formalize their approach to occupational safety and health through a safety program will reduce the numbers and severity of accidents and injuries. The degree of this reduction is directly proportional to the resources devoted to the formal program.

In a speech presented to the U.S. House of Representatives Committee on Small Business in September 1999, Charles N. Jeffress, Assistant Secretary of Labor for Occupational Safety and Health, reported that "...thirty-two states have some form of safety and health program provision, though few are as comprehensive as OSHA's draft proposed rule. Four states (Alaska, California, Hawaii, and Washington) have mandated comprehensive programs that have core elements similar to those in OSHA's draft proposal, that cover businesses of all sizes within the state, and for which at least five years of data are available. In those four states, injury and illness rates fell by nearly 18% over the five years after implementation, in comparison with national rates over the same period. Several other states have studied the effectiveness of their own programs and found that average workers' compensation costs were reduced by as much as 20% per year, and that these benefits were even greater several years later when the program had matured."

Mr. Jeffress summarized by stating that "experience with safety and health programs demonstrates that systematic, common-sense efforts to protect workers have a direct impact on workplace injury and illness rates and on compliance with existing worker protections ... [and] the common-sense advantages provided by safety and health programs will reduce these injuries, illnesses, fatalities, and associated workers' compensation costs, bringing a clear new benefit to the many establishments that have yet to establish such programs."

Again, to add credence to these OSHA recommendations, it has been the experience of the authors during hundreds of loss control consultative visits that, time and time again, those organizations that incorporated some or all of these guidelines enjoyed a more productive and less injurious work environment.

A sample safety program is included in Appendix A and is provided to assist in the establishment of a written program or as a comparison to what your organization currently has in effect. Either way, the important point to remember is that written safety programs should describe the organization's approach to occupational safety and health *and* the organization should follow the program. In addition, the contents of the program should be inclusive of all workplace hazards and the methods of control. It is a "living document" that must evolve with the hazards. Keep it current with periodic reviews.

2.3   Safety committees

2.3.1   Necessity

Communication is critical to the success of any organization, so much so that the nation's colleges and universities are graduating students with Bachelor of Science degrees in organizational communications. In addition, organization effectiveness consultants immediately focus on the communication systems internal and external to a business entity, as this characteristic of a business is either the cause of or the cure for its trials and tribulations. What a company does has less significance in comparison to how it does it, and communication is a critical component of how companies conduct their business.

Perhaps the most beneficial communications component of safety management is the safety committee. In essence, it is a conduit of information (facts, perceptions, ideas, and recommendations) that supplements the traditional managerial authority lines typically in existence for all other managerial functions — personnel, production, resource, etc. It adds a dimension of quality to any safety program, and for large organizations it is essential.

Consider the challenges confronted by organizations whose operational interests or geographical locations are many and varied. The usefulness of a safety committee becomes obvious. One such organization was visited by the author for a loss control consultation. The following recommendation was the primary corrective action offered as a solution to the compartmentalized, disrupted information flow internal to that company. The recommendation demonstrates how useful a safety committee can be and also how it can function.

> Institute a formal agency safety committee with a written description of its goals, authority, responsibility, and membership. The frequency of its convening should be no less often than once a month. All agency programs and outlying facilities should have a representative in attendance. The agency director should

also participate. Discussion should focus on safety-
related issues, but understand that many person-
nel/operational/logistical topics that impact upon
safety will surface, and that is good. This committee
may very well function as a critical focal point for
many issues that are currently being addressed separately. All changes in business practice — whether it's
new employees, new policies, new procedures, new
equipment, or new facilities — have the potential to
create new occupational safety and health hazards.
Only when all parties that are affected by these changes convene and exchange experiences/recommenda-
tions will effective and established feedback occur. In
addition, cost savings might be realized by implementing and then sharing the analysis of the agency's
loss history on a monthly basis. Program "A" may
have a solution to an injurious situation currently
faced by Program "B." The monthly safety committee
meeting provides the forum for this discussion. Considering the numerical, operational, and geographical
diversity of the agency, it becomes necessary that a
committee be established. One individual will be partially successful in managing the safety program information collection and analysis, but the preventive
actions that may be a result of this information processing are guaranteed to fail without organizational
involvement and support. The key to this organizational involvement and support is the safety committee and leadership by example.

2.3.2   Considerations

2.3.2.1   Purpose

Too often committees have the reputation of wasting time and accomplishing
nothing, and too often that reputation is valid. Without a specified and
clearly defined purpose, the safety committee will degenerate into doing
whatever the attendees want to do on that particular date. Formalizing the
committee's purpose in writing is effective, but sustained adherence to the
purpose is what establishes the committee as a solid forum. Typical objectives of a safety committee may include

- Discussion of and input to safety policy
- Review of safety training needs and evaluation of training completed
- Explanation of new safety standards, codes, or laws enacted by reg-
  ulatory agencies
- Analysis of accident and injury data

- Review of accident and near-miss accident investigations
- Presentation of recommendations or items of concern

2.3.2.2 Membership

A common mistake is to put too many of the wrong people on the committee. What's the right number? It depends. We've all experienced meetings with too many people in attendance. Communication was stifled or few of the agenda items could be addressed in the allotted time.

Members of the safety committee should provide a representative sampling of the interests contained in the business (production, resource management, maintenance, personnel management) and it should include a cross section of the hierarchy of the business. If membership is homogenous, with only managers, then the committee's actions will reflect that characteristic. Consequently, managers, supervisors, and line employees should all be represented. Some recommend no more than ten members, but this decision must be made in light of the situation in your company. For large organizations, a successful approach has been to create department safety committees that, in turn, have a representative on the company safety committee.

Committee members should be volunteers who possess the requisite skills and attitudes necessary to facilitate the committee process. This is not to say that they are all "yes men," but that they can contribute to the committee's effectiveness as opposed to sabotaging every effort.

Some committee members, like subject matter experts (SMEs), may attend meetings only when their expertise is required. This provides flexibility to the committee. SMEs include industrial hygienists, computer information systems specialists, transportation director, and company attorney.

Another source of committee membership is outside the company. The loss control representative and/or the workers' compensation claims adjuster from the insurance provider can be excellent choices. These individuals have vast and specialized experience to share with the committee, besides their vested interest.

2.3.2.3 Budget

An unfunded committee is destined for an early demise; in other words, no money, no credibility. The expenditures are not large, but they exist. Expect to incur cost for training of committee members. Seminars in "people skills" or specific safety topics that the committee is addressing can quickly enhance the committee's results and they can strengthen the committee members' dedication to the effort. Travel funds may be required for those organizations that are geographically dispersed — committee meetings can be held at a different location each month. No one department should bear this expense; charge it against the committee. Finally, the committee will use supplies and possibly need some equipment. All of these will cost something — plan for it and request a budget.

2.3.2.4 Conduct

Someone must assume control of the committee meetings, and the safety director is the likely candidate for this role. However, if someone is not designated as the committee director (or leader, facilitator, chair, or manager), then an informal leader will emerge. At that point, the objectives of the committee will most likely be disregarded and hidden agendas will surface.

The committee chair should publish the agenda for the next meeting and distribute it to committee members well in advance of the meetings. This procedure adds credibility and structure to the proceedings. It also provides a subtle reminder to those who have data to present that they are expected to be prepared.

Make the meetings routine (same date each month, same time of day, same sequence of events, etc.), but rotate their location. The rotation enables all committee members to get away from their locale and to see other aspects of the company. It is surprising how many employees never get the chance to see the "big picture."

Proceedings of the meetings should be recorded, in writing, and made available for the company's employees to review. There are no secrets to hide. Nothing confidential should be discussed in the safety committee meetings — it is not the appropriate forum. Meeting minutes should include the results of action items previously assigned to committee members, some of which may have been suggested by an employee. When employees see their suggestions are taken seriously and fairly considered, the committee gains the employees' respect and confidence. It shows the committee is open to input, acts on that input, and reports the results. This does not mean the committee must incorporate all suggestions, but that all suggestions are accepted and addressed. This recognition validates the worth of the individual who submitted the input and in turn that validation strengthens the human resource component of the business. And what an important resource employees are!

2.4   Personnel management

2.4.1   Employee turnover

Without exception, every business in this country is challenged by employee turnover. Typically, it has been due to seasonal fluctuation, layoff-recall, retirement, and anticipated attrition of the work force. The contributing factors to current employee turnover rates are numerous, and many believe the number is unprecedented in our national history. Regardless, employee turnover can undermine the productivity of a company and significantly increase the organization's occupational injury rate.

2.4.1.1   Negative effects of turnover

Some of these effects can be easily assessed by their financial drain on a company's resources; others are more insidious and elusive. But the bottom line is that *they all cost something and the business must pay.* The nonfinancial costs include morale, business image, supervisory attention, and training.

- *Morale* — When a vacancy exists, typically a task is not performed or it is performed by someone in addition to his or her regularly assigned duties. Either way, those employees who remain at work become frustrated and eventually their morale will decline. They too may then seek employment elsewhere.
- *Business Image* — Be honest. What is your impression of a company that is constantly hiring individuals but has no apparent business expansion underway? Have you ever stopped at a fast-food restaurant that was advertising for employment opportunities on its marquee? That's when you realized that "Help Wanted" in the fast-food industry could be translated into "Slow Service." One disgruntled employee with low morale can instantly negate thousands of dollars of advertisement investment by simply displaying his or her negative feelings to a customer. What's at stake is the business image; it takes months or even years to establish a good one but only seconds to destroy it.
- *Supervisory Attention* — An understaffed business is overworked, and the quality of service or products will decline if supervision is not increased. The situation frustrates the supervisors, who are likely to cry out, "Let's just hire more people!" That's the apparent solution. But what impact does the presence of a new employee have on a supervisor? It consumes even more of the supervisor's time and attention, *and it should.* National data indicates that most occupational accidents involve those employees who have been with the company for less than 180 days. These employees are, then, the highest risk portion of the work force. Consequently, even if the company can remain fully staffed, the supervisor's frustration continues because of the increased chance of accidents associated with new employees. Turnover is detrimental.
- *Training Costs* — Training is an area that can be very expensive. Assuming that the mandatory, annual OSHA training is conducted, and that technically specific training is conducted, and that internal company procedural training is conducted, and that all this training can be quantified financially, figure what your organization's turnover rate is and then increase your annual training costs by that figure. In essence, you are training individuals for employment elsewhere.

2.4.1.2 *It's happening — but why?*

Granted, the fact that the national economy has sustained expansion for several years is the primary reason for turbulence in the work force, but there are many other contributing factors. Identifying these may facilitate the preparation of effective retention efforts. It's often said that identification of the problem is much more difficult than solving the problem. Factors contributing to employee turnover include employee dissatisfaction, the "greener grass" myth, social change, demographics, and multiple opportunities.

2.4.1.2.1 Employee dissatisfaction.   Employee dissatisfaction is a huge category. Some common causes of it are listed below.

- Incomplete job information for new hires
- Ineffective interviewing techniques
- Lack of training
- No promotional opportunity
- Low wages and few benefits
- Poor working conditions
- Inadequate supervision and organizational structure
- Domestic problems

2.4.1.2.2 Greener grass myth.   Things always seem to look better from a distance, and employment opportunities are no different. The type of industry, location of the facility, coworkers, benefits, hours of operation, travel destinations, managerial approach, etc. all may have some attractive quality. Unfortunately, too often these qualities are no different than those that existed in the previous place of employment. When the employee takes a close look, he or she sees the grass really wasn't any greener, so the process begins anew.

2.4.1.2.3 Social change.   Psychologists and sociologists are having a field day studying social change in all aspects of life, but its impact on employee turnover is real. The term "instant gratification generation" usually creates a negative image. Has this attitude permeated our work force? As a nation, are we impatient for solutions or gratification? Do we want employment satisfaction as quickly as we obtain electronic resolution (cellular phones, computers, e-mail, television sitcoms, microwave cooking, e-commerce, etc.)?

And what about loyalty? This could be another of those disappearing national virtues. But loyalty is a two-way street. Downsizing, mergers, business reorganizations, buyouts, and layoffs can create the perception of the company's "disloyalty" toward its employees. Individuals may use those examples as justification for "job hopping," describing their actions as career enhancement, individual betterment, etc. The result is the same — employee turnover.

2.4.1.2.4 Demographics.   As the baby boomers mature into the autumn of their years, they continue to exert strong, although sometimes subtle, influences on the economy. In addition to the unprecedented creation and consumption of goods and the consequential employment opportunities, the passage of that "bulge" of workers through the national occupational infrastructure may be leaving a trail of unfillable jobs in its wake. Add to this situation the fact that many of the boomers achieved financial independence and the reinvestment of their capital created even more employment opportunities.

2.4.1.2.5 Multiple opportunities. There's really no need to belabor this point. Why remain at Company X this morning when I can work at Company Y this afternoon? It is definitely a job seeker's market. The classified ad section in all newspapers validates this fact daily. These announcements also reinforce the desire for instant gratification, flaunt the greener grass myth, and even cause the individual to reflect on his or her current occupation. Sometimes these announcements are very tempting, especially if the employee has any dissatisfaction with his or her current employment.

2.4.1.3 Relationship with workers' compensation

Employee turnover has become a vicious cycle and represents one of the few disadvantages of a continuously booming economy. But it also underscores the importance of human resources and their retention. It's interesting to note that trucks, machines, and additional warehousing can be procured or constructed within weeks, but how long does it take to recruit, identify, employ, train, and retain quality individuals?

Minimizing employee turnover will positively affect the productivity of the business. It will also contain the occupational injury rate because the high-risk portion of your work force is reduced.

2.4.2 Employee retention

As already stated, the segment of the work force that sustains the most injuries is the employees who have been with the company for less than 180 days, the high turnover portion of the work force. If these individuals are increasing the cost of workers' compensation insurance, then it stands to reason that attention should be focused on reducing the size of this population. The logical solution is employee retention — the conscious and deliberate effort to retain quality individuals on the company payroll. Stated otherwise, it's the proactive methods utilized by successful organizations to stop the drain of company profits caused by excessive employee turnover.

Perhaps this is easier said than done, but in reality some companies *are not conceding* to the national trend of high turnover and *are preserving their work force.* Employee retention is reflected in their occupational injury rates; they are sustaining fewer accidents. It's not mystical. There is a direct correlation between low employee turnover and fewer accidents/injuries. Consider the following strategies of employee retention.

2.4.2.1 Sources of replacements

Granted, vacancies exist and always will. But how does your company seek replacements? Does it really matter how they do it? Yes, it does.

What level of confidence do you have when a stranger walks in and asks for an employment application? Have you ever contracted the services of a personnel placement agency? They usually prescreen applicants and match their knowledge, skills, and abilities to the vacant jobs. Don't forget your best recruiters — your current employees. They will make referrals, and a good employee will typically be a good referral. A common replacement method is advertisements in newspapers; for some this works best.

The point is that multiple sources of hiring are available and each has its advantages and disadvantages. None of them is best in every circumstance. Realize this fact and include all sources in your arsenal of personnel replacement options.

2.4.2.2   Management

Human resources are the most important, most expensive, and most challenging business resources. How do you manage yours?

All individuals should be interviewed *prior* to hiring — not *after*! Does your company have established interview procedures? Who interviews? Where? When? What questions are asked? And then, who makes the decision?

Once hired, an employee should be provided a written job description. This document enables that individual and his or her supervisor to clearly understand exactly what is supposed to be accomplished and to what standard. Everyone needs guidance, employees and supervisors alike. By putting the job description in writing, everyone, employee and supervisor alike, knows what is expected. Periodic performance appraisals should be completed for all employees. Everyone needs constructive feedback, positive as well as negative. These appraisals also serve as a management tool for promotion, retention, reassignment, and training decisions.

If, and when, an employee departs, an exit interview will gather valuable information (quite often perceptions) about the company. People will speak more freely upon departure. Some of this information is critically important for improving retention. Typical questions for exit interviews include why is the employee leaving, what was the most appealing thing about the company, and what would the employee change.

Communication with employees must be established upon application and then maintained throughout their period of employment. Hiring/exit interviews, written job descriptions, and periodic performance appraisals are all examples of personnel management communications.

2.4.2.3   Training

Did you realize that quality training not only raises the skill levels of employees but also retains employees? One of the most often cited reasons for employee departure is dissatisfaction with the company's training program.

Traditional training programs include new employee training and professional development, both of which are designed to improve the employee's productivity. Another dimension of training that has been proven to increase productivity and to increase retention is "life skills" training. This type of training has evolved over the last few years as a retention measure. Topics include financial and retirement planning, stress management, creating work/family balance, and time management. Of course, these skills also have a positive impact on the employee's contribution to the company by reducing the possibility of workplace violence or substance abuse.

For younger, aspiring employees, company-sponsored training opportunities are quite appealing.

2.4.2.4   Benefits

A list of benefits is limited only by management's imagination. Some benefits are common, while others are unusual. Some are expensive and others are free. Generally none appeals to every employee; therefore, several incentives and benefits should be available to retain your good employees. Typical benefits include

- Competitive wages
- Medical/dental insurance
- 401-K savings plan
- Paid vacations
- Company stock options
- Flex time
- Relaxed dress code
- Child care
- Employee assistance programs
- Wellness programs

For many employees, some of the above might not hold much interest — for example, an employee without children won't be able to take advantage of daycare — but the *availability* of these benefits gives them the feeling that they are valued as an individual. Moreover, simple and basic human kindness and caring can foster loyalty, such as knowing their birthday and giving them the day off — with pay — or maintaining a bulletin board with news articles describing their children's accomplishments.

2.4.2.5   Purifying the force

Not all employees qualify for retention, so identify those who don't and let them go! The structured interview process and the periodic performance appraisals will purify the force. In addition, these individuals are typically more prone to injuries and once injured are more difficult to return to work. Let them go! However, employees should not be released simply because they have sustained an injury. This practice will quickly invite a wrongful termination lawsuit and then infect the morale and confidence of other employees. Purification of the workforce is a continuous process that begins with recruitment and includes periodic performance appraisals and counseling statements. The individual's performance is measured against a standard — and that implies that a written job description exists. Terminations should not be whimsical. They should be based upon failure to meet a prescribed standard of performance.

2.4.2.6 Advantages

Minimizing employee turnover through the deliberate efforts to retain quality individuals will have a positive effect on the productivity of your business. It will also contain the occupational injury rate because the high-risk portion of your work force has been reduced. Remember that it is less expensive in the long run to retain than it is to recruit.

2.5   Training

As previously noted, training has such fundamental importance to a comprehensive occupational safety and health program that it is one of the four sections of OSHA's recommended guidelines. Training does not happen of its own accord, and once is not enough; it should be managed and continuous. Consider the following statements:

- Training can be a tremendous expense for an organization.
- Effective training programs are productivity multipliers.
- Ineffective training programs are consumers.
- All companies deserve and expect a return on their training dollar investments.
- Safety training is a subset of the company's overall training program.
- Safety training reduces the frequency and severity of accidents and injuries.

There is risk involved with training — investment risk. Time, personnel, capital, and physical plant are all resources that must be diverted towards the task of training. An unorganized and purposeless management of this investment will undermine its success and jeopardize the opportunity for future training. The individual responsible for training should be aware of this risk and must invest those resources wisely.

2.5.1   OSHA training guidelines

OSHA's approach to the challenge of training is described below. The model is designed to be one that even the owner of a business with very few employees can use without having to hire a professional trainer or purchase expensive training materials. Using this model, employers or supervisors can develop and administer safety and health training programs that address problems specific to their own business, fulfill the learning needs of their own employees, and strengthen the overall safety and health program of the workplace. These training guidelines follow a model that consists of the following steps:

- determining if training is needed
- identifying training needs
- identifying goals and objectives
- developing learning activities
- conducting the training
- evaluating program effectiveness
- improving the program

This approach is applicable to occupational safety and health training as well as all other types of training (new employee training, proficiency enhancement training, professional development training, life skills training). Training management is training management, so you can also use this approach when addressing training as a means of employee retention.

2.5.1.1 Definitions

Prior to describing OSHA's training guidelines, it's necessary to have a common definition of three key concepts — training, performance, motivation — that underlie all training programs. It is equally important that the relationship among these concepts be understood.

In general, *training* refers to instruction and practice for acquiring skills and knowledge of rules, concepts, or attitudes necessary to function effectively in specified situations. With regard to occupational safety and health, training can consist of instruction in hazard recognition and control measures, learning safe work practices and proper use of personal protective equipment, and acquiring knowledge of emergency procedures and preventive actions. Training can also provide workers with ways to obtain added information about potential hazards and their control; for example, they could gain skills to assume a more active role in implementing hazard control programs or to effect organizational changes that would enhance worksite protection.

Performance, on the other hand, represents observable actions or behaviors reflecting the knowledge or skill acquired from training to meet a task demand. With regard to occupational safety and health, performance can mean signs of complying with safe work practices, using protective equipment as prescribed, demonstrating increased awareness of hazards by reporting unsafe conditions to prompt corrective efforts, and executing emergency procedures should such events occur.

Finally, *motivation* refers to processes or conditions that can energize and direct a person's behavior in a way intended to gain rewards or satisfy needs. Setting goals for performance coincident with learning objectives and use of feedback to note progress have motivational value. With regard to occupational safety and health, motivation can mean one's readiness to adopt or exhibit safe behaviors, take precautions, or carry out self-protective actions as instructed. Bonuses, prizes, or special recognition can act as motivational incentives or rewards in eliciting as well as reinforcing these behaviors when they are displayed.

Realize, though, that knowledge or skills acquired in training may not always result in improved performance in actual work situations. When they don't, it may indicate a lack of suitable motivation, training content that does not fit job demands (inadequate training objectives), or dissimilarity or conflicts between the instruction or practice in the training and actual job conditions. Herein lies the challenge faced by the trainer and the training manager.

2.5.1.2   Determining if training is needed

The first step in the training process is a basic one: determine whether training can solve a problem. Whenever employees are not performing their jobs properly, it is often assumed that training will bring them up to standard. However, it is possible that other actions (such as hazard abatement or the implementation of engineering controls) would enable employees to perform their jobs properly.

Ideally, safety and health training should be provided before problems or accidents occur. Such training covers both general safety and health rules and work procedures and is repeated if an accident or near-miss incident occurs.

Problems that can be addressed effectively by training include those that arise from lack of knowledge of a work process, unfamiliarity with equipment, or incorrect execution of a task. Training is less effective (but still can be used) for problems arising from an employee's lack of motivation or lack of attention to the job. Whatever its purpose, training is most effective when designed in relation to the goals of the employer's total safety and health program.

2.5.1.3   Identifying training needs

If the problem is one that can be solved, in whole or in part, by training, then the next step is to determine what training is needed. To make that decision, it is necessary to identify what the employee is expected to do and in what ways, if any, the employee's performance is deficient. This information can be obtained by conducting a job analysis that pinpoints what an employee needs to know in order to perform a job.

When designing a new training program or preparing to instruct an employee in an unfamiliar procedure or system, a job analysis can be developed by examining engineering data on new equipment or the safety data sheets on unfamiliar substances. The content of the specific federal or state OSHA standards applicable to a business can also provide direction in developing training content. Another option is to conduct a Job Hazard Analysis. This procedure studies and records each step of a job, identifying existing or potential hazards and determining the best way to perform the job in order to reduce or eliminate the risks. Information obtained from a Job Hazard Analysis can become content for the training activity.

If an employee's learning needs can be met by revising an existing training program rather than developing a new one, or if the employer already has some knowledge of the process or system to be used, appropriate training content can be developed through such means as

- Using company accident and injury records to identify how accidents occur and what can be done to prevent them from recurring.
- Requesting employees to provide, in writing and in their own words, descriptions of their jobs. These should include the tasks performed and the tools, materials, and equipment used.
- Observing employees at the worksite as they perform tasks, asking about the work, and recording their answers.
- Examining similar training programs offered by other companies in the same industry or obtaining suggestions from organizations such as the National Safety Council (which can provide information on Job Hazard Analysis), the Bureau of Labor Statistics, OSHA-approved state programs, OSHA full-service area offices, OSHA-funded state consultation programs, or the OSHA Office of Training and Education.

The employees themselves can provide valuable information on the training they need. Safety and health hazards can be identified by asking employees if anything about their jobs frightens them, if they have had any near-miss incidents, if they feel they are taking risks, or if they believe that their jobs involve hazardous operations or substances.

There are regulatory requirements that dictate training (upon hire, when work conditions change, periodic, etc.). At first these regulations can be confusing to the novice occupational safety and health manager, but with a little experience they become common knowledge. For example, once the safety manager fully implements the bloodborne pathogen standard, forklift standard, or hazard communication standard, the requirement for periodic training is a component of that program. The who, what, when, where, why, and how regarding training is plainly stated in the standard.

Once the kind of training that is needed has been determined, it is equally important to determine what kind of training is *not* needed. Employees should be made aware of all the steps involved in a task or procedure, but training should focus on where improved performance is needed. This avoids unnecessary training and tailors training to meet the needs of the employees.

2.5.1.4   Identifying goals and objectives

After identifying training needs, employers can prepare objectives for the training. Clearly stated instructional objectives will tell employers what they want their employees to do, what to do better, or what to stop doing.

Learning objectives do not necessarily have to be written, but in order for the training to be as successful as possible, clear and measurable objectives should be thought out before the training begins. For an objective to be effective it should identify as precisely as possible what the individuals will do to demonstrate what they have learned or that the objective has been reached. They should also describe the important conditions under which the individual will demonstrate competence and define what constitutes acceptable performance.

Using specific, action-oriented language, instructional objectives should describe the preferred practice or skill and its observable behavior. For example, rather than using the statement "The employee will understand how to use a respirator" as an instructional objective, it would be better to say "The employee will be able to describe how a respirator works and when it should be used." Objectives are most effective when worded in sufficient detail that other qualified persons can recognize when the desired behavior is exhibited.

Conducting training without a predetermined performance objective amounts to nothing more than training for the sake of training, and that is a waste of resources. The training may have been very enjoyable, with great visual aids and a popular instructor, but if the design of that training lacked measurable objectives, then it will most likely be ineffective.

2.5.1.5   Developing learning activities

Learning activities enable employees to demonstrate that they have acquired the desired skills and knowledge. To ensure that employees transfer the skills or knowledge from the learning activity to the job, the learning situation should simulate the actual job as closely as possible. Thus, employers may want to arrange the objectives and activities in the order in which the tasks are performed on the job. For example, if an employee must learn the beginning processes of using a machine, the sequence might be (1) check that the power source is connected, (2) ensure that the safety devices are in place and are operative, and (3) know when and how to throw the switch.

A few factors will help to determine the type of learning activity to be incorporated into the training. One aspect is the training resources available to the employer. Can a group training program that uses an outside trainer and film be organized, or should the employer personally train the employees on a one-to-one basis? Another factor is the kind of skills or knowledge to be learned. Is the learning oriented toward physical skills (such as the use of special tools) or toward mental processes and attitudes? The training activity can be group-oriented, with lectures, role-play, and demonstrations, or designed for the individual, as with self-paced instruction.

The methods and materials for the learning activity are also various. The employer may want to use charts, diagrams, manuals, slides, films, overhead transparencies, videotapes, audio tapes, computer-based training modules, or simply blackboard and chalk, or any combination of these and other instructional aids. Whatever the method of instruction, the learning activities should be developed in such a way that the employees can clearly demonstrate that they have acquired the desired skills or knowledge. It is important to note that the opportunity for the employee to interact with the instructor is especially important whenever computer-based training modules are incorporated. Their use as a stand-alone training program without any instructor assistance is not recommended, but when used with an instructor they can definitely enhance a program.

2.5.1.6 Conducting the training

With the completion of the steps outlined above, the employer is ready to begin conducting the training. The training should be presented so that its organization and purpose are as clear as possible to the employees. Employers or supervisors should (1) provide overviews of the material to be learned; (2) relate, whenever possible, the new information or skills to the employees' goals, interests, or experience; and (3) reinforce what the employees learned by summarizing the program's objectives and the key points of information covered. These steps will also assist employers in presenting the training in a clear, unambiguous manner.

In addition to organizing the content, employers must also develop the structure and format of the training. The content developed for the program, the nature of the workplace or other training site, and the resources available for training will help employers determine for themselves the frequency of training activities, the length of the sessions, the instructional techniques, and the individuals best qualified to present the information. Figure 2.1 graphically displays many of these considerations.

In order to be motivated to pay attention and learn the material that the employer or supervisor is presenting, employees must be convinced of the importance and relevance of the material. Among the ways of developing motivation are (1) explaining the goals and objectives of instruction; (2) relating the training to the interests, skills, and experiences of the employees; (3) outlining the main points to be presented during the training session; and (4) pointing out the benefits of training (e.g., the employee will be better informed, more skilled, and thus more valuable both on the job and on the labor market; or the employee will, if he or she applies the skills and knowledge learned, be able to work at reduced risk).

An effective training program allows employees to participate in the training process and to practice their skills or knowledge. This will help to ensure that they are learning the required knowledge or skills and permit correction if necessary. Employees can become involved in the training process by participating in discussions, asking questions, contributing their

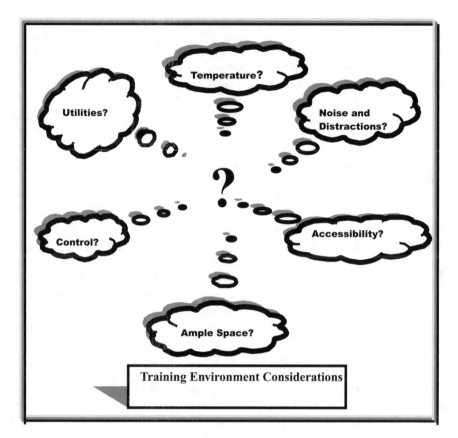

Figure 2.1

knowledge and expertise, learning through hands-on experiences and role-playing exercises.

2.5.1.7   Evaluating program effectiveness

To see if the training program is accomplishing its goals, an evaluation of the training can be valuable. Training should have, as one of its critical components, a method of measuring its effectiveness. A plan for evaluating the training session should be developed when the course objectives and content are developed. It should not be delayed until the training has been completed. Evaluation will help employers or supervisors determine the amount of learning achieved and whether an employee's performance has improved on the job. Methods of evaluating training include

- Student opinion — questionnaires or informal discussions with employees can help employers determine the relevance and appropriateness of the training program.

- Supervisors' observations — supervisors are in a good position to observe an employee's performance both before and after the training and note improvements or changes.
- Workplace improvements — the ultimate success of a training program may be changes throughout the workplace that result in reduced injury or accident rates.

However it is conducted, an evaluation of training can give employers the information necessary to decide whether or not the employees achieved the desired results and whether the training session should be offered again at some future date.

2.5.1.8 Improving the program

After evaluation, if it is clear that the training did not give the employees the level of knowledge and skill that was expected, then it may be necessary to revise the training program or provide periodic retraining. At this point, asking questions of employees and of those who conducted the training may be of some help. Among the questions that could be asked are: (1) Were parts of the content already known and therefore unnecessary? (2) What material was confusing or distracting? (3) Was anything missing from the program? (4) What did the employees learn, and what did they fail to learn?

It may be necessary to repeat steps in the trainings planning and implementation. As the program is evaluated, the employer should ask

- If a job analysis was conducted, was it accurate?
- Was any critical feature of the job overlooked?
- Were the important gaps in knowledge and skill included?
- Was material already known by the employees intentionally omitted?
- Were the instructional objectives presented clearly and concretely?
- Did the objectives state the level of acceptable performance that was expected of employees?
- Did the learning activities simulate the actual job?
- Were the learning activities appropriate for the kinds of knowledge and skills required on the job?
- When the training was presented, was the organization of the material and its meaning made clear?
- Were the employees motivated to learn?
- Were the employees allowed to participate actively in the training process?
- Was the employer's evaluation of the program thorough?

A critical examination of the steps in the training process will help employers to determine where course revision is necessary. Assuming that quality instruction accomplished the goal is a false assumption. Training must be evaluated.

2.5.2   Matching training to employees

While all employees are entitled to know as much as possible about the safety and health hazards to which they are exposed, and while employers should attempt to provide all relevant information and instruction to all employees, the resources for such an effort frequently are not, or are not believed to be, available. Thus, employers are often faced with the problem of deciding who is in the greatest need of information and instruction. Prioritization must occur.

One way to differentiate between employees who have priority needs for training and those who do not is to identify employee populations that are at higher levels of risk. The nature of the work will provide an indication that such groups should receive priority for information on occupational safety and health risks.

2.5.2.1   Identifying employees at risk

One method of identifying employee populations at high levels of occupational risk (and thus in greater need of safety and health training) is to pinpoint hazardous occupations. Even within industries that are hazardous in general, some jobs have greater risk than others. In other cases the hazard of an occupation is influenced by the conditions under which it is performed, such as noise, heat or cold, or safety or health hazards in the surrounding area. In these situations, employees should be trained not only on how to perform their job safely but also on how to operate within a hazardous environment.

A second method of identifying employee populations at high levels of risk is to examine the incidence of accidents and injuries, both within the company and within the industry. If employees in certain occupational categories are experiencing higher accident and injury rates than other employees, training may be one way to reduce that rate. In addition, thorough accident investigations can identify not only specific employees who could benefit from training but also identify company-wide training needs.

Research has identified the following variables as being related to a disproportionate share of injuries and illnesses at the worksite on the part of employees:

- age of the employee (younger employees have higher incident rates)
- length of time on the job (new employees have higher incident rates)
- size of the company (generally, medium-sized companies have higher incidence rates than smaller or larger firms)
- type of work performed (incident and severity rates vary significantly by Standard Industrial Code)
- use of hazardous substances (by Standard Industrial Code)

These variables should be considered when identifying employee groups for training in occupational safety and health. This information is readily available to help employers identify which employees should receive safety and health information, education, and training, and who should receive it before others. Employers can request assistance in obtaining information by contacting such organizations as OSHA area offices, the Bureau of Labor Statistics, OSHA-approved state programs, state onsite consultation programs, the OSHA Office of Training and Education, or local safety councils. Trade organizations are also good sources of information.

2.5.2.2   Training employees at risk

Determining the content of training for employee populations at higher levels of risk is similar to determining what any employee needs to know, but more emphasis is placed on the requirements of the job and the possibility of injury. One useful tool for determining training content from job requirements is the Job Hazard Analysis described earlier. This procedure examines each step of a job, identifies existing or potential hazards, and determines the best way to perform the job in order to reduce or eliminate the hazards. Its key elements are

- job description
- job location
- key steps (preferably in the order in which they are performed)
- tools, machines, and materials used
- actual and potential safety and health hazards associated with these key job steps
- safe and healthful practices, apparel, and equipment required for each job step.

Material Safety Data Sheets (MSDS) can also provide information for training employees in the safe use of chemicals. These data sheets, developed by chemical manufacturers and importers, are supplied with manufacturing or construction materials and describe the ingredients of a product, its hazards, protective equipment to be used, safe handling procedures, and emergency first aid responses. The information contained in these sheets can help employers identify employees in need of training (i.e., workers handling substances described in the sheets) and train employees in the safe use of the substances. MSDS are generally available from suppliers, manufacturers of the substance, or larger employers who use the substance on a regular basis, or employers or trade associations can develop them. Several Internet sites also provide them. It should be noted that the OSHA Hazard Communication Standard specifically requires that employers conduct training on chemical use at their particular company, and MSDS are useful for the development and conduct of this training.

2.5.3   Summary

Occupational injuries result from either unsafe actions or unsafe conditions. Training may be the solution to minimizing their occurrence, but the training must be a *managed* effort. OSHA's model addresses the questions of who should be trained, on what topics they should be trained, and for what purposes they should be trained. It also helps employers determine how effective the program has been and enables them to identify employees who are in greatest need of education and training. The model is general enough to be used in any area of occupational safety and health training (or production training) and allows employers to determine for themselves the content and format of training.

It is interesting to note that in 1998 the National Institute for Occupational Safety and Health (NIOSH) commissioned a study to assess occupational safety and health training. The researchers found that more than 100 OSHA standards for hazard control in the workplace contain requirements for training aimed at reducing risk factors for injury or disease; others limit certain jobs to persons deemed competent by virtue of special training. A review of 80 reports, from 1980 to 1996, on training aimed at reducing risk of work-related injury and disease provided overwhelming evidence of the merits of training in increasing workers' knowledge of job hazards and in effecting safer work practices and other positive actions in a wide array of worksites. Reports from select surveys and investigations of worker injuries and workplace fatalities were also assessed, with many implicating lack of training as a contributing factor to the mishaps. In still other studies, workplace training devoted to first aid instruction showed linkage to reduced worker injury rates, suggesting that even this kind of training has benefits to job safety overall. In addition, variables such as size of training group, length/frequency of training, manner of instruction, and trainer credentials were each shown to be significant determinants to the training process. Equally important were factors unrelated to training, such as goal setting, feedback, and motivational incentives, along with managerial actions to promote the transfer of learning to the job site.

The literature review offered much direct and indirect evidence to show the benefits of training in ensuring safe and healthful working conditions. Especially supportive is data from training intervention reports addressing major workplace hazards in a wide array of work situations. Findings were nearly unanimous in showing how training can attain objectives such as increased hazard awareness among the work groups at risk, knowledge of and adoption of safe work practices, and other positive actions that can reduce the risk and improve workplace safety. Other data suggest that inadequate or lack of required safety training may have contributed to events in which workers were injured or killed. While affirming the benefits of occupational safety and health training, some limitations in the data sources were also noted bearing on the merits of current OSHA training rules in reducing work-related injury and disease. Even with these shortcomings, evidence that occupational safety and health training can reduce risks from workplace hazards remains strong. Indeed, the issue is not so much whether it is worthwhile but what factors both within and beyond the training process can produce the greatest possible impact.

2.6 Incentive awards

Depending upon the day of the week and whom you ask, safety incentive awards are one of the more effective means of reinforcing safe behavior and reducing occupational injuries, or they represent an opportunity for employees, supervisors, and managers to falsely report injuries — thus masking the true incident rate of an organization and precluding the investigation of those accidents. The truth depends on several factors within and specific to each company.

Safety incentive awards are just that — an incentive that reinforces desired behaviors and encourages the continuance of that behavior. It is a positive, affirmative motivational tool that is intended to decrease, or eliminate, the occurrence of workplace injuries.

2.6.1 Considerations

The success of safety incentive awards programs is predicated upon numerous considerations. Similar to training programs, successful incentive awards programs don't just happen — they must be managed.

- Incentive awards programs are not a replacement for a safety and health program — they are another dimension of the existing program. If your organization does not currently have a comprehensive safety and health program, then creating one should be your higher priority.
- Simple programs are manageable. An intricate, burdensome, administrative nightmare will cause the program manager to dread the duty and eventually shirk the responsibility. Keep it simple.
- Programs are typically managed unilaterally, but they are collateral in participation. Get everyone involved. Without upper management's support, the program will stall; without supervisors' support, the program will lack credibility; and without the employees' support, the program will not attain its objective. This support is inclusive of both suggestions and participation in the formulation, maintenance, and revision of the program. For example, what safe behaviors does the organization want to encourage and how will the organization reward those behaviors? Has management committed to the funding of the program? The appropriate forum for this management could very well be the organization's safety committee. If you don't have one, now you have a reason to establish one.

- As for all goals, program goals should be understandable, realistic, obtainable, and measurable. What is somewhat unique to incentive awards program goals is that the time frame should be specific because the reward must be closely associated with the behavior. Otherwise, the positive benefit of reinforcing the desired behavior is lessened.
- Programs must be reviewed and modified. Typical questions to be asked periodically include whether the incentives are desired by the employees and if the program has targeted the intended audience. Providing employees with awards that mean little or nothing to them is not likely to have a positive effect on the incident rate. Don't let the program grow stale or become an entitlement. Review it and change it to meet the company's needs. Keep the program invigorating and motivational.
- Incentive awards programs must be visible. Don't keep it a secret. Obviously, the more personnel who assist with the project, the more publicity it will receive. But there needs to be a planned, systemically executed program awareness campaign. Announce its beginning. Publicize participants' progress. Communicate directly with contenders, encouraging their efforts. Congratulate those who attain the goals.

2.6.2 Challenges

Realize that there are pitfalls associated with safety incentive awards programs and that they will challenge the program manager. The solution lies within the management of the program. Common myths and attitudes about safety incentives, such as the following, can undermine a program:

- Employees should not be rewarded for doing their job; that's why they receive a paycheck.
- Incentive awards will only create false reporting of injuries, thus precluding the opportunity to truly know what workplace hazards have yet to be controlled.
- Incentive awards will allow negative peer pressure to fester.
- Employees will perceive incentive awards as an entitlement, especially when the program is changed or terminated.
- No one wants to get hurt, so there is no point in rewarding someone for not getting hurt.
- The expense of the program far exceeds the return on investment.

This list could go on and on, and there is merit in these beliefs because all too often safety incentive programs are not well designed, marketed, and subsequently managed. If the considerations listed above are incorporated into the development and maintenance of an incentive awards program, the myths will be dispelled. Quantifiable, positive effects on the organization's incident rate and workers' compensation premium will validate the program's value. In addition, employee morale, retention, and productivity will increase because a basic human need — recognition — is being met.

2.7   Summary

The successful accomplishment of a task begins with a plan. The avoidance of workers' compensation claims is also predicated upon a plan, commonly known as a safety and health program. The components of a safety and health program include the owners and employees of the organization, their attitudes towards safety, and a formal document that describes how the organization will approach its activities to prevent accidents and injuries. If any of these three components is deficient, then the results will reflect a commensurately deficient loss history.

The focus of this chapter has centered on the establishment of a formal safety and health program. It is assumed that with the adoption, implementation, and continuous modification of such a program, the safety attitudes in the organization will elevate and all employees will, by the very nature of their employment, participate. The formal program is the road map and the employees and management are the navigators. It is a living document that reflects the current approach to safe production operations. When, and if, those operations change, then the program is evaluated and changed in accordance. Dust-covered reams of written safety policies, procedures, and programs that are neatly filed on someone's shelf are worthless. The value of a safety program lies in its execution. Formalizing your organization's safety program can reduce the frequency and severity of occupational injuries, thus avoiding workers' compensation claims.

chapter three

Identification and control of workplace hazards

Contents

3.1   Introduction

Although accident investigations and subsequent reviews are a part of the worksite analysis referred to in OSHA's guidelines, they represent a *reactive* approach to loss control, one that occurs after the accident has caused damage, injury, or business interruption. A successful safety and health program is *proactive*. Not only does it include accident investigations and review, but it also seeks to identify workplace hazards and to eliminate or control hazards prior to accidents. For that reason, this chapter first discusses proactive methods of hazard identification and control and then briefly discusses the more common, but reactive, accident investigations.

     Proactive hazard identification and subsequent control is best accomplished with a combination of comprehensive hazard identification, routine site safety and health inspections, and employee reports of hazards. In reality, unfortunately, the proactive portion of worksite analysis is often conducted either with little coordinated effort or without conscious thought of the process. Either way, the true benefits of the analysis usually are not realized. For example, the specific work procedures involved with the operation of a machine are typically described to a new employee prior to his unsupervised work release and considered an "orientation." Safety engineers, however, consider such a description of operations a Job Hazard Analysis (JHA); a formal, written description of the procedures, risks, and control measures surrounding the conduct of any job. The difference between the commonplace new employee orientation, which is seldom in writing, and the formal JHA is the systemic approach utilized in the analysis and the subsequent corrective actions taken by management to prevent an accident. If an organization formalizes its worksite analysis, then corrective actions can occur prior to placing an employee at risk.

     The relationship between accidents and the components involved in occupational activities is graphically displayed in Figure 3.1. It is notable that accidents can result from either unsafe acts or unsafe conditions, and either of these can be introduced at any point during the continuum of operations. One "check" most often will not be sufficient hazard identification and control.

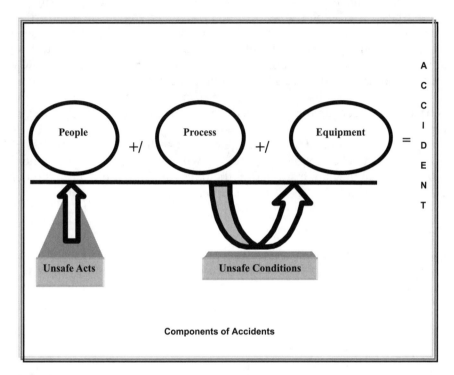

Figure 3.1

There are numerous approaches to identifying and then quantifying the hazards that exist in the workplace. Described below is a formal and systemic approach that incorporates both employees and management. It establishes a baseline and then monitors for any evolving or uncontrolled hazards.

3.2   Hazard identification methods

A complete hazard inventory is necessary in order to develop a program of prevention and control. Common shortcomings in this effort include inventories limited in scope or time. This assessment should be designed from a systems standpoint — all aspects of the business (people, equipment, procedures, environment) should be included and the assessment should be continuous. Identifying only some of the workplace hazards or failing to understand the fact that hazards change is not sufficient. With these objectives, therefore, a comprehensive hazard identification program should include wall-to-wall assessments, change analyses, and job hazard analyses.

3.2.1   Wall-to-wall assessments

This is the most basic of all the tools used to establish the inventory of hazards and potential hazards at your worksite. Experts from outside the worksite who have knowledge that is both broad and deep perform this survey best. That expertise should include safety engineering, industrial hygiene, and, in most cases, occupational medicine. After a baseline has been established, only periodic comprehensive surveys need to be done to take advantage of new information in the experts' communities about newly understood hazards or means of controlling them. Examples of this change in hazards and the subsequent evolution of experts' knowledge include the concept of indoor air quality, bloodborne pathogens exposure, and even ergonomics.

The incorporation of a safety engineer in the physical design process of new construction or renovation projects will often identify the potential hazards before they are instituted in the physical plant. More discussion of this topic is included in the subsequent section pertinent to hazard control.

3.2.2   Change analysis

Change analysis implies that to solve a problem or anticipate its occurrence, an investigator must look for changes from the norm. Each time there is a change of facilities, equipment, processes, or materials in your workplace, they should be analyzed for hazards before they are introduced. This way, not only is harm to workers avoided, but the extra expense of retrofitting controls after installation and use is also avoided. As its name implies, this technique emphasizes change.

And just what is the norm? Usually we recognize it only after it has been altered. For example, note the disrupted work pattern, cycle, attention span, and even quality in an organization soon after a significant change has occurred — change in supervisors, new machine or vehicle in the production process, accident, department realignment, building expansion or renovation, personnel layoffs, new employees, or chemical substitution. These are typical changes whose effects are first noted by the production manager, but the astute safety manager realizes that the potential for accidents and injuries has also increased due to these changes. This increase in risk should be anticipated.

3.2.3   Job hazard analysis

The basic form of this analysis, which is useful at every type of worksite, is also known as the Job Safety Analysis (JSA). It divides a job into tasks and steps, analyzes the hazards potential in each step, and then provides a method of prevention or control to reduce exposure. JSA involves both the employee (affected individual) and an observer (supervisor or safety specialist). The result is a written description of the task (not the job) that lists the sequential steps, embedded hazards, and methods of hazard control associated with that particular task.

There is both an art and science of a JSA. The observer should have a general appreciation for the occupational task and should be adept in the JSA process itself. A common pitfall that undermines the quality and subsequent utility of a JSA is that its completion is relegated to a new hire or an intern as "busy work." Another mistake is failing to explain to the employee the JSA procedure and then to solicit that employee's input regarding the steps involved in the task. It is critical that the employee understand that the JSA observation is not a critique of his or her performance and abilities, but rather a description of the task. Without communicating this to the employee before the JSA commences, the observer inadvertently corrupts the data because the employee may deviate from normal operational procedures due to the suspicion of being personally evaluated.

After obtaining a sequence of the steps involved in the task, the observer should then examine each step to determine the presence of any hazards or to speculate as to whether hazards might be generated. The OSHA booklet *Job Hazard Analysis* offers the following questions as samples of what the observer might consider during the JSA.

- Are there materials on the floor that could trip a worker?
- Is lighting adequate?
- Are there any live electrical hazards at the job site?
- Are any chemical, physical, biological, or radiation hazards associated with the job or likely to develop?
- Are tools — including hand tools, machines, and equipment — in need of repair?
- Is there excessive noise in the work area, hindering worker communication or causing hearing loss?
- Are job procedures known and are they followed or modified?
- Are emergency exits clearly marked?
- Are trucks or motorized vehicles properly equipped with brakes, overhead guards, backup signals, horns, steering gear, and identification, as necessary?
- Are all employees who are operating vehicles and equipment properly trained and authorized?
- Have any employees complained of headaches, breathing problems, dizziness, or strong odors?
- Is ventilation adequate, especially in confined or enclosed spaces?
- Have tests been made for oxygen deficiency and toxic fumes in confined spaces before entry?
- Are workstations and tools designed to prevent back and wrist injuries?
- Are employees trained in the event of a fire, explosion, or toxic gas release?
- Is the worker wearing personal protective clothing and equipment, including safety harnesses that are appropriate for the job?

- Are work positions, machinery, pits or holes, and hazardous operations adequately guarded?
- Are lockout procedures used for machinery deactivation during maintenance procedures?
- Is the worker wearing clothing or jewelry that could get caught in the machinery or otherwise cause a hazard?
- Are there fixed objects that may cause injury, such as sharp machine edges?
- Is the flow of work improperly organized (e.g., is the worker required to make movements that are too rapid)?
- Can the worker get caught in or between machine parts?
- Can the worker be injured by reaching over moving machinery parts or materials?
- Is the worker at any time in an off-balance position?
- Is the worker positioned to the machine in a way that is potentially dangerous?
- Is the worker required to make movements that could lead to or cause hand or foot injuries or strain from lifting — the hazards of repetitive motions?
- Can the worker be struck by an object or lean against or strike a machine part or object?
- Can the worker fall from one level to another?
- Can the worker be injured from lifting or pulling objects or from carrying heavy objects?
- Do environmental hazards — dust, chemicals, radiation, welding rays, heat, or excessive noise — result from the performance of the job?

Once the task has been observed and the hazards or potential hazards identified, the next challenge is to determine if alternate procedures can eliminate those hazards. If not, then physical changes should be considered, such as equipment redesign, machine guarding, personal protective equipment, or ventilation as a means of containing the hazard. If hazards remain even after all these control and containment methods have been implemented, then the frequency of exposure should be minimized. Remember, risk is equal to the combination of exposure and hazards. If you cannot totally eliminate or contain one, then you should address the other.

Multiple benefits can be derived from compiling an inventory of current JSAs. Primarily, occupational hazards associated with specific tasks have been identified, quantified, eliminated, or contained. Secondary benefits are also generated. Job safety analyses are excellent documents for new employee training. In conjunction with demonstration, there is not a better way to orient an individual to a task. Employees are also empowered during the JSA process because they are allowed to participate. Their input has merit and they will appreciate the opportunity to contribute to a safer workplace.

Depending upon the size and diversity of a workplace, the process of performing and updating job safety analyses can become overwhelming and even costly. When this is the case, the savvy safety manager will prioritize those jobs that incur the most accidents or injuries and start at the top. If the results of this limited safety intervention of high-injury jobs show reduced incident rates or improved productivity, this can be a strong argument for increased departmental funding to expand the JSA scope. Another cohort who would probably support the safety manager's request for funding of 100% JSA completion could be the manager who has responsibility for new employee training. As mentioned earlier, JSAs are great training outlines that are produced in-house. These documents complement safety, training, and production. The goal is to analyze each task in every job and periodically update those analyses.

3.2.4   Routine safety and health inspections

Humans are creatures of habit, and these habits can be good or bad. Inspections of the workplace should be a habit. They should be scheduled events, conducted by someone who is held accountable, facilitated through the use of checklists. The guesswork has been eliminated because the who, what, when, where, why, and how have become routine.

Checklists are prolific and their availability is virtually instant with the advent of the Internet. Portions of the checklists posted on the OSHA web site are included in Appendix B and are provided as a sampling. Other web sites, publications, and organizations offer different formats and different topical areas. The point is that checklists are a tool to be used in the identification of workplace hazards and that no two workplaces are identical; therefore, there is not one universally applicable checklist.

The individual who conducts the inspection should either have subject matter expertise (be proficient in the specific operational area under inspection) or be trained in the procedure of identifying workplace hazards, or both. Numerous and diverse sources of inspectors are readily available to assist in the process, as displayed in Figure 3.2. Each can contribute to the identification, elimination, or control of workplace hazards, which ultimately leads to a decreased number of occupational injuries. Most inspectors are only a phone call away.

3.2.4.1   Internal inspectors

General site inspections should be performed by personnel at the worksite who have been trained in recognizing hazards that have a tendency to slip by the controls designed to reduce them. Inspectors should also keep an eye out for hazards that may not have been identified in the comprehensive survey or through any other means of establishing the original hazard inventory. Typically, these inspections are conducted with checklists that have been developed locally, acquired from a regulatory agency, or provided by an equipment manufacturer. The value of these inspections is their frequency (sometimes each shift) and the fact that the individual conducting

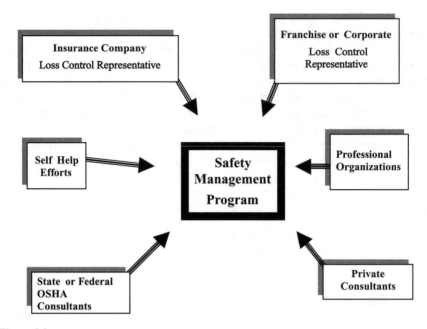

Figure 3.2

the inspection is quite often the individual who is either exposed to the hazard or supervises those who are.

Inspections, in and of themselves, serve little purpose unless corrective action is initiated and then verified for effectiveness. How many times during an accident investigation have we noted that an employee had reported equipment deficiencies on his daily checklist but that no corrective action had been taken? Sometimes the corrective action was taken and assumed to be sufficient, but later, after an accident, it becomes clear that the corrective action itself created another hazard. In this situation a checklist was utilized but the second and third steps (correction and verification) were ignored. The situation, however, is much worse when checklists are not utilized at all.

Also realize that by using every employee as an extra pair of eyes to spot hazards, you can greatly increase your speed and reliability in finding and correcting problems before any harm is done. Who has more at stake for the identification and control of hazards than the individual employee who is at risk? In this approach checklists are not used. Instead, the employee's individual expertise and experience are the standards for judgment. Such reports should provide employees with anonymity and should be "hassle free." Otherwise, few employees will take the time to alert management of the hazards. Employees also need to see timely and appropriate responses to their reports to encourage them to continue their input. Such responses also provide employees with evidence of management's commitment to the safety and health program. Most managers will recognize this concept as the suggestion awards program, or as a component of the safety incentives program.

3.2.4.2   OSHA consultations

There are two methods of utilizing the services of government occupational safety and health regulatory agencies — request a free consultation or wait until they appear at your door to conduct an unannounced inspection. The results are similar, but the processes are not. Of all the companies visited by the authors that have partaken of the free service, each has irrefutably benefited from the event, and most have requested subsequent visits. However, others believe that requesting an OSHA consultation is equivalent to dealing with the devil. Lay those fears aside. If your intent is to reduce the occurrence of occupational injury and contain the cost of workers' compensation, then OSHA consultations are the way to go.

3.2.4.2.1   Background.   Employers who want assistance implementing effective safety and health programs to prevent injuries and illnesses resulting from hazardous workplace conditions and practices can get it from a free consultation service largely funded by the Occupational Safety and Health Administration. The consultation program not only addresses immediate problems but also offers advice and help in maintaining continued effective protection. For the short term, consultants help employers identify and correct specific hazards. For the long term, they provide guidance in establishing or improving an effective safety and health program and offer training and education for employers and employees. The service is given chiefly at the worksite, but limited services may be provided away from it.

Primarily targeted for smaller businesses in higher-hazard industries or with especially hazardous operations, the safety and health consultation program is completely separate from the inspection effort. The service is also confidential. Your name and firm and any information about your workplace, plus any unsafe or unhealthful working conditions that the consultant uncovers, will not be reported routinely to the OSHA inspection staff. In addition, no citations are issued or penalties imposed as a result of a consultation. Your only obligations are to allow the consultant to confer with employees in the course of the hazard survey and to correct any imminent dangers and other serious job safety and health hazards in a timely manner. You make these commitments before the consultant's visit. Consultation is a cooperative approach to solving safety and health problems in the workplace. As a voluntary activity, it is neither automatic nor expected. You must request it.

3.2.4.2.2   Benefits.   An obvious benefit of using a safety consultation is that if you know the hazards in your workplace and ways to remedy them, you are in a better position to comply with job safety and health requirements. The more you know about the safety and health aspects of your company's operations and ways to improve them, the better you can manage your company in general.

Another benefit is that when a consultant helps establish or strengthen a workplace safety and health program, safety and health activities become routine considerations rather than crisis-oriented responses. Moreover, under this program you may be exempted from general schedule OSHA enforcement inspections for one year when you have a complete examination of your workplace, correct all identified hazards, post a notice of their correction, and institute the core elements of an effective safety and health program (described in Chapter 2). Using this consultation service, employers can find out about potential hazards at their worksites, improve their occupational safety and health management systems, and ultimately reduce their incidence rate and workers' compensation premiums — much to be gained at no cost.

3.2.4.2.3 Procedures. In a consultation, the consultant is both student and teacher. The consultant studies each workplace and the employer's safety and health program. Based on this analysis, the consultant then applies his or her expertise to the specific problems and unique operations of the workplace and instructs the employer on these applications.

Consultation can go beyond the usual physical survey of the workplace for violations of federal or state OSHA standards. It's important to note here that compliance solely with regulatory standards does not guarantee the absence of workplace injuries. Therefore, the consultant may also point out work practices not yet covered by the standards but that are likely to cause illness or injury and then advise and assist the employer in correcting them. He or she may propose other measures directed toward improving an organization's injury and illness experience. These measures may include using self-inspection, emphasizing supervisory responsibility in promoting safety, identifying safety and health training needs, alerting workers to hazards, employing labor-management safety and health committees, and holding regular safety and health meetings with workers. Comprehensive consultation services include the following:

- Appraisal of all mechanical and environmental hazards and physical work practices,
- Appraisal of the present job safety and health program or the establishment of one,
- Conference with management on findings,
- Written report of recommendations and agreements,
- Training and assistance with implementing recommendations, and
- A follow-up to assure that any required corrections are made.

3.2.4.2.4 Getting started. The consultation process is initiated with your request, which may be a telephone call, letter, or personal contact. Some services, such as a safety and health review of proposed or new production processes, may be conducted at locations away from the employer's worksite.

When you request on-site services, your request will be prioritized according to the nature of your workplace and any existing backlog of requests. A consultant assigned to your request will contact you to establish a visit date based on the priority assigned to your request, your work schedule, and the time needed for the consultant to prepare adequately to serve you. The consultant may encourage you to include within the scope of your request all working conditions at the worksite and your entire safety and health program. You retain the option, however, to limit the consultation visit to a discussion of fewer, more specific, problems. If the consultant observes hazards that are outside the scope of the request, he or she is obligated to notify you of their presence.

3.2.4.2.5 Opening conference. Upon arrival at your worksite for a scheduled visit, the consultant will briefly review his or her role during the visit and will want to review with you your safety and health program. The consultant will explain the relationship between on-site consultation and OSHA enforcement activity and further explain your obligation to protect employees in the event that serious hazardous conditions are identified. Also, he or she will explain that employee participation is encouraged during the consultation process.

3.2.4.2.6 Walkthrough. During the walkthrough of the worksite, you and the consultant will examine conditions in your workplace. The consultant will identify any specific hazards and provide advice and assistance in establishing or improving your safety and health program and in correcting any hazardous conditions identified. At your request, assistance may also include education and training for you, your supervisory personnel, and other employees.

OSHA strongly encourages, but does not require, worker participation in the walkthrough. Better-informed and alert workers can more easily work with you to identify and correct potential injury and illness hazards. At a minimum, the consultant must be able to talk freely with workers during the walkthrough to help identify and judge the nature and extent of specific hazards and, where requested, to evaluate your safety and health program.

The consultant will study those areas upon which all agreed at the beginning. As mentioned earlier, he or she will also offer advice and assistance on other safety or health hazards that might not be covered by current federal or state OSHA standards but that still pose safety or health risks to your employees.

In a complete review of a company's operation, the consultant looks for mechanical and physical hazards by examining the structural condition of the building, the condition of the floors and stairs, and the exits and fire protection equipment. During the tour of the workplace, he or she reviews the layout for adequate space in aisles and between machines, checks equipment such as forklifts, and examines storage conditions. Control of electrical hazards and machine guards is also considered.

The consultant checks the controls used to limit worker exposure to environmental hazards such as toxic substances and corrosives, especially air contaminants. He or she checks whether all necessary technical and personal protective equipment is available and functioning properly. Also, the consultant notes any problems workers may encounter from exposure to noise, vibration, extreme temperatures, lighting, or other environmental factors and offers the techniques commonly used for the elimination or control of any hazards.

Work practices, including the use, care, and maintenance of hand tools and portable power tools, as well as general housekeeping, are of interest to the consultant. He or she wants to talk with you and your workers about items such as job training, supervision, safety and health orientation and procedures, and the maintenance and repair of equipment. In addition, the consultant wants to know about any ongoing safety and health program your organization has developed. If your organization does not have a program or you would like to make improvements, at your request the consultant offers advice and technical assistance on establishing a program or improving it. Management and worker attitude toward safety and health is considered in this analysis, as well as current injury and illness data. The consultant also needs to know how you and your employees communicate about safety and health as well as to be aware of any in-plant safety and health inspection programs.

3.2.4.2.7 Closing conference. Following the walkthrough, the consultant has a closing conference with you. This session offers the consultant an opportunity to discuss measures that are already effective and any practices that warrant improvement. During this time, you and the consultant can discuss problems, possible solutions, and time frames for eliminating or controlling any hazards identified during the walkthrough.

In rare instances, the consultant may find an "imminently dangerous" situation during the walkthrough. In such situations, an employer must take immediate action to protect all affected workers. If the consultant finds a hazard that is considered to be a "serious violation" under OSHA criteria, he or she will work with you to develop a mutually acceptable plan and schedule to eliminate or control that hazard. During this time, OSHA encourages you to advise all affected employees of the hazards and to notify them when the hazards are corrected. Consultants offer general approaches and options as well as technical assistance on the correction of hazards when they have the expertise. As necessary, consultants will recommend other sources for specialized technical help.

The consultant may also offer suggestions for establishing, modifying, or adding to the company's safety and health program to make them more effective. Such suggestions could include worker training, changing work practices, methods for holding supervisors and employees accountable for safety and health, and various methods of promoting safety and health.

3.2.4.2.8 Hazard correction and program assistance. After the closing conference, the consultant sends you a written report explaining the findings and confirming any correction periods agreed upon. The report may also include suggested means or approaches for eliminating or controlling hazards as well as recommendations for making your safety and health program effective. You are, of course, free to contact private consultants for additional assistance at any time.

Ultimately, OSHA does require completed action on serious hazards so that each consultation visit achieves its objective — effective worker protection. If an employer fails or refuses to eliminate or control an identified serious hazard or any imminent danger in accordance with the plan or any extensions granted, the situation would be referred to an OSHA enforcement office for review and action, as appropriate. However, these are rare occurrences.

3.2.4.2.9 Summary. The consultation program provides several benefits for you as an employer. On-site consultants:

- Help you recognize hazards in your workplace,
- Suggest approaches or options for solving a safety or health problem,
- Identify sources of help available to you if you need further assistance,
- Provide you with a written report that summarizes these findings,
- Assist you in developing or maintaining an effective safety and health program,
- Offer training and education for you and your employees at your workplace and, in some cases, away from the site, and
- Under specific circumstances, recommend you for recognition by OSHA and a 1-year exclusion from general schedule enforcement inspections.

Consultants do not

- Issue citations or propose penalties for violations of federal or state OSHA standards,
- Routinely report possible violations to OSHA enforcement staff except for unabated serious conditions, or
- Guarantee that any workplace will "pass" a federal or state OSHA inspection.

3.2.4.3 Loss control specialists

Essentially, there are two types of loss control specialists — those who work for your insurance carrier and those who freelance through the auspices of consulting agencies. Their levels of expertise, professional credibility, and methods of operation are very similar, but the cost of their services is different. Most workers' compensation insurance carriers maintain a safety department staffed with loss control specialists. This department is responsible for assisting the insured company in establishing and maintaining the occupational safety and health of its employees as well as protecting the interests of the insurance carrier. The services of the loss control specialists are included in the insurance premium. Loss control specialists who represent consulting agencies are equally beneficial, but they charge for their services. Both are held to professional standards of conduct and should keep the occupational safety and health of the employees paramount in the execution of their duties.

3.2.4.3.1 Private consultants. Loss control consultants are available virtually everywhere. Deciding upon a specific agency is no different than selecting any other contractor. Typical questions to ask include

- What services are provided?
- What do the services cost?
- Do the consultants have references?
- Are the consultants responsive to the individual needs of your company?
- Will their product benefit your operation?

Most loss control specialists can be located in the Yellow Pages of the phone directory or in the classified section of trade publications. A simple search on the Internet can also provide hundreds of leads within seconds.

The decision to use a private loss control consultant remains with the business owner. If your workers' compensation insurance carrier is not providing this service, or if you desire loss control consultation in greater depth, then a private consultant is definitely warranted. Many of the insureds visited by the authors also have contracts with private loss control consultants, but most often the primary reason for the contracts is to secure periodic safety training to comply with regulatory guidance. Evidence of their assistance quickly surfaces, however, and it is all beneficial.

3.2.4.3.2 Insurance carrier loss control representatives. As previously stated, most insurance companies maintain a loss control or safety department staffed with degreed or credentialed safety professionals. Their focus is less on regulatory compliance and more on hazard identification and control. As such, the objective of incorporating their assistance is to reduce the opportunity for occupational injury. They may offer exactly the same recommendations as a private consultant in terms of training or use of personal protective equipment, or they may delve deeply into a specific area of occupational risk that has resulted in a loss history replete with injuries. Their objective is to identify the hazards, quantify the risk, and then recommend control measures so as to reduce and ultimately eliminate the personal injury loss incidence.

It is important to note that an organization can be in complete regulatory compliance (although this is rare and exists only temporarily) and still sustain occupational injuries. Likewise, the same organization can be out of compliance and not sustain occupational injuries. The diagram in Figure 3.3 presents this relationship and offers some explanation. The point is not to undermine the value and importance of regulatory compliance, but rather to emphasize that a "fixation" on regulatory compliance often lulls an organization into believing that all is well. The transportation industry is a classic example of this point. Often, trucking companies will devote 99% of their safety efforts to compliance with Department of Transportation regulations; however, some of the companies that receive laudatory ratings during their DOT audits are also incurring employee injuries as a result of drivers falling off the trailers, slipping on the steps as they enter and exit the truck cabs, or cargo being hit by when they open the trailer doors. The loss control representatives from your workers' compensation insurance provider will pursue the full range of occupational hazards associated with your company, generally not limiting themselves to regulation-related hazards, and offer recommendations for their control — if you let them!

The astute business owner has already read between the lines. Loss control representatives basically serve two masters — the insurer and the insured. They serve the insured by assisting with the safety program and striving to eliminate occupational injuries. They serve the insurer by accurately describing the insured's risk and then attempting to assist the insured with containing losses. For the most part, the loss control representative is the only individual from the worker's compensation insurance provider that ever visits an insured and gathers a total business description (interviews, observations, walkthrough inspections, perceptions, etc.), so an accurate description of the business's safety situation is important to better enable the underwriter to properly assess premiums. If loss control representatives are denied entrance to the business or not afforded full opportunity to accomplish their objective while on the site, then the insured does not benefit from the loss control representative's expertise in reducing risk and loss. Moreover, the insurer is likely to become leery of the insured and may be conservative in granting any premium credits. In such a case, mutual trust and confidence are not allowed to flourish. And for those whose premiums are significant, discounts of 5% and 10% are significant.

Consultants and insurance company loss control representatives offer recommendations, items to consider for implementation, but they are not absolutes or dictates. Instead, they are offered as a potential solution for a perceived risk. Why, then, do some business owners refuse to incorporate loss control representatives, a powerful resource that can immediately reduce their incident rate, in their overall safety management program?

Just as the OSHA consultant summarizes inspection findings in a written report, so too should the loss control representative or consultant. The quality of the report is indicative of the quality of the inspection. When shopping for consultants or assessing the capabilities of your workers' compensation

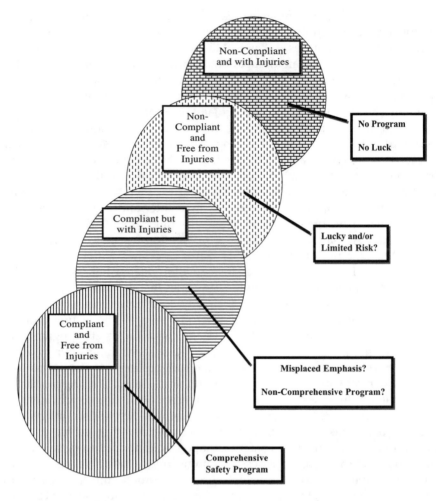

Figure 3.3

carrier's loss control department, ask to see a sample of their product. The following comments were extracted from such a report. The inspection did not cost anything (it came with the workers' compensation policy) except about two hours of the business owner's time and attention. What a bargain!

I. Loss experience summary
During the period of 01/01/99 through 12/31/99, this company submitted 7 compensable workers' compensation claims, the reserved cost of which is $57,665. As such, this company has a loss ratio of 199.1 and a loss frequency rate of .24 per $1,000 of earned premium

during the above-mentioned period. It is notable that this period does *not* correspond with the policy period.

The following is a summary of the loss history of this company during the period of January 1, 1999 through December 31, 1999.

Month	Claims
January	1
February	2
March	0
April	1
May	0
June	2
July	0
August	2
September	0
October	0
November	0
December	0

A majority of these injuries were purely ergonomic or resulted in soft tissue damage as the result of impact from falls.

II. Synopsis of visit
An initial LEAP visit was conducted November 5, 1999 with Mr. Smith, the production manager. As such, the opportunity to expeditiously discuss current business operations and the company safety and health program during this visit was facilitated.

Two specific areas of interest were explored — modified duty/return-to-work and medical management of work place injuries. This interest was generated from the presence of three claims that resulted in extensive compensation expenditures as well as medical expenses. Mr. Smith explained that the company does, indeed, have the philosophy of assertively attempting to accommodate injured employees with modified duty, and that injured workers are escorted to the medical treatment facilities. Both of these actions are proactive measures in containing the costs associated with worker's compensation claims.

The issue of high employee turnover and its negative impact on safety was also discussed.

A wall-to-wall observation of the facility, during operational hours, was conducted. It was noted that the employees in the paint booth were wearing air-purifying respirators and some discussion regarding a respirator program management ensued. Mr. Smith requested whatever additional literature pertinent to respirators the Loss Control Department may have on file. A recent addition to the physical plant was also toured. Mr. Smith explained that the facility allowed employees to fabricate towers without having to work at elevated heights.

III. Actions pending/recommendations

Quarterly Loss Control Assistance Visits will be conducted henceforth for Fictitious Manufacturing. Written summaries of these visits will be forwarded to Mr. Smith and Mr. Jones (Chief Financial Officer).

As requested, attached is an additional source of information regarding the recently revised Respiratory Standard. Further information may be obtained online by accessing the Internet site (www.safetyinfo.com).

The following recommendations are offered for consideration/action.

- Continue to intervene in the early stages of medical attention following workplace injuries. The end results are the minimization of the employee's absence from work; an increase in positive psychological and emotional feelings experienced by the employee that enhance the physical healing process (he/she feels wanted and needed by the organization); the maximization of the potential productivity of an experienced, trained employee (eliminates the need to hire temporary help); and containment of the compensation costs associated with the claim. This is a win/win scenario, but management must be willing and prepared to immediately become involved in the medical treatment process. Another benefit of this approach is the steady increase of an employee's loyalty to the company — and this positive affirmation will eventually whittle away at the high turnover rate currently experienced.

- Review the safety incentives offered to employees for no-lost-time injuries as a means of lowering the turnover rate. If the incentives can be modified (cash awards as opposed to tangibles), perhaps employees will not only work safer but also remain on the payroll long enough to collect. Statistically, the initial period of employment for an individual in any organization is the most dangerous. A disproportionate number of injuries occur with individuals who have been employed for 180 days or less. The minimization of turnover reduces the size of the population that is at higher risk, not to mention the tremendous advantage it offers for productivity. The national average estimate for the cost of turnover for each employee ranges between $10K and $20K. Needless to say, a reduction in the turnover rate will reap benefits throughout the company's operational areas. The safety incentives may be a *currently funded effort* that, with modification, could assist with this reduction.
- Hold all supervisors accountable for the timely and accurate completion of accident reports. They represent the solution to implementing proposed changes. If they are exempt from the process, either by design or default, then the information and action chain is *broken*.
- Fully incorporate Internet information sites into your repertoire of sources of solutions to occupational safety and health challenges. A daily visit to the National Institute of Occupational Safety and Health (www.cdc.gov/niosh) and the Occupational Safety and Health Administration (www.osha.gov) will be a productive investment of your time. Also, join the list-serv managed by the University of Vermont (www.esf.uvm.edu/uvmsafetyinfo/safetywelcome.html). Thousands of safety professionals visit this site daily, posting information, asking questions, or providing answers. Your challenges may have already been addressed by another manufacturing organization. Review the ergonomic information available at the web sites www.advancedergonomics.com,

www.ctdnews.com, www.osha.gov, and www.inter-face-analysis.com/ergoworld/ to determine if any recommended work practices have application to the physically challenging duties of Fictitious Manufacturing Company employees.

- The best effort to reduce the incident rate of ergonomic injuries may lie in training and education. All too often our level of safety awareness becomes minimal over time, particularly in regard to common tasks. Many of the strains incurred by the employees are sustained during common tasks. Contact a local health provider (physical therapist, chiropractor, sports medicine trainer) and explore the opportunity for them to present a class on lifting safety. Perhaps they would oblige without any fee in return for the publicity. Or a local or regional safety consultant could be contracted to prepare and present a *site-specific* series of ergonomic awareness classes. If these prove unsuccessful, select the appropriate training material available through the Loss Control Department and present it in an employee safety meeting. The important point is to discuss lifting safety, soon and often.

- The frequency rate of ergonomic injuries may be reduced by more selectively screening new hires during the application process. Preexisting injuries may be sufficient evidence to preclude assigning an individual to a high-risk job without adequate protection. Protective measures range from redesigning the job to providing appropriate training; hence, the recommendation above is made.

- In addition, those employees who apparently get hurt more often than others may need more training or more supervision. They definitely need attention. Whatever corrective actions are taken should be documented, either as a record of training or as a counseling record. An organization need not surrender itself to a "klutz." Supervisors should intervene. Even consider the institution of an annual employee appraisal system that would hold both the supervisor and the employee accountable for their performance of duty.

3.2.4.4   *Franchise or corporate safety officers*

If you have a relationship with a corporation or franchise, seek assistance from its loss control representative. Most likely you will not get an on-site visit, but your mailbox will fill with numerous facility inspection checklists, a recommended annual training program (preferably with supporting lesson plans), safety posters and signs, and periodic newsletters.

Don't despair; usually these items are worthwhile and *free*. Sort through the materials, and use what you can. To reinforce this point, the following true story is shared. While conducting an annual loss control consultative visit with an insured, the author inquired whether the company maintained any formal, written loss control programs or an inclusive safety program. The business owner shrugged his shoulders and indicated that he thought that about $2,000 had been provided to a regional occupational safety and health consulting agency to produce a customized safety program, but he really didn't know what, if anything, had become of the expenditure. He had not seen the product. We asked the same question of the office manager and she replied by pointing to a bound, cardboard box that was sitting on the floor next to the trash can. Partly to satisfy my curiosity as to what a $2,000 safety program contained and to acquaint the business owner with his new program, I asked if I could open the box. The customized program was top notch, detailed and impressive, but it was worthless while it remained in the box! The primary comment included on the loss control summary was, "Identify an individual to be responsible for the company safety program. Ensure that he/she is empowered to execute the program. The documents prepared by XYZ Consultants, Inc. are an excellent foundation for your program, but without follow-through, the program exists in name only."

3.2.4.5   *Self help*

Some Internet web sites offer numerous capabilities to identify and control workplace hazards. Of particular utility are downloadable software programs from the OSHA web site known as "safety advisors." This concept is new, rapidly improving and expanding, and offers much potential. Essentially, the software is developed under OSHA's endorsement and is designed to enable businesses and others to answer a few simple questions and receive reliable answers on how OSHA regulations apply to their unique worksite. Expert Advisors are provided free and can be downloaded and run on local personal computers (PCs). The Expert Advisors combine the expertise of OSHA safety and health professionals, including epidemiologists, risk assessors, and attorneys, into a single source of expert help. As of the writing of this text, the selection of advisors includes programs concerned with asbestos, confined spaces, fire safety, cadmium, hazard awareness, lead, lockout/tagout, respiratory protection, healthcare facilities, safety and health programs, and even the computation of the financial impact of lost workdays. Three of these areas — hazard awareness, fire safety, and safety and health programs — have direct application to the general identification and control of workplace hazards, while the others become more specific. Each of these three areas is described in more detail below.

- Hazard Awareness Advisor — designed to assist in the identification of hazards in general industry workplaces; asks about your workplace and asks follow-up questions based on your answers. It then provides a customized report about possible hazards and related OSHA rules.
- Fire Safety Advisor — addresses OSHA's general industry standards for fire safety and emergency evacuation and for fire fighting, fire suppression, and fire detection systems and equipment. Once installed on your PC, it asks about building and business policies and practices. It then asks follow-up questions based on your answers and prepares the guidance and customized plans you need.
- Safety and Health Program Advisor — also known as a Program Evaluation Assistant; helps you review and evaluate key aspects of your safety and health program if you have one. If you do not have one, it can help you think about elements of a good program. It is straightforward and very easy to use.

All of these can be quickly downloaded to your PC and incorporated into your workplace hazard assessment process. They are not a complete solution, but when used in conjunction with other assessment tools and techniques they greatly supplement the process.

Another source of help available in most regional colleges or universities that offer occupational safety and health classes is the instructors and the students who may be willing to voluntarily conduct a workplace hazard assessment as a class project. Understand, though, that the product may not be of the highest quality; however, it is free and represents a good start point for your subsequent concentrated efforts. Acquaintance with the instructors also affords the ability to network with credible safety consultants because degree-producing educational institutions represent a focal point for those involved in the profession. If such an institution is close at hand, contact the department chair and inquire. There's no harm in asking.

3.2.4.6 Synergistic effects

An additional advantage of having external inspectors visit your facility, regardless of who they are or what organization they represent, is the opportunity to observe their inspection techniques. In so doing, you can enhance your abilities to identify workplace hazards on your own.

There is both an art and a science to conducting physical inspections. Total reliance upon a checklist is rather narrow. Only after sufficient experience is acquired does the inspector fully develop the art of inspection — that ability to sense or anticipate subsequent conditions predicated upon observed conditions.

Accompanying a visitor through your facility while he or she conducts a physical inspection presents this opportunity, but only if someone accompanies the inspector and makes a sincere effort to understand the process. The benefits are synergistic – you profit from the inspector's observation of your facility and you improve your abilities to self-inspect.

3.2.5 Accident investigations and near-miss reviews

The primary tool used for determining the cause of an accident is an accident investigation. Implemented as a stand-alone loss control tool, accident investigations will likely offer only limited results. However, combined with other management initiatives, accident investigations are one of the most effective methods of reducing work-related injuries. Therefore, accident investigations and near-miss reviews are integral elements of an effective injury prevention program. In fact, these assessment/analytical tools are so integral that Chapter 8 is devoted entirely to their discussion.

The investigation or review of an accident or near-miss incident is a retrospective activity, but one that can lead to prospective benefits. It will do nothing for the injured employee or the company's incident rate — that's history. What it can do, though, is preclude the same, or similar, injuries from recurring. Although the accident has occurred and subsequent injury resulted, some advantages, as depicted in Figure 3.4, may be gained from the situation, provided a structured, deliberate investigation is accomplished. Include these reviews and investigations in your approach to identifying and controlling workplace hazards. The information gleaned from them can leverage the effectiveness of your injury prevention and reduction efforts.

Benefits Derived from Accident Investigations

- Identifies root causes of accidents
- Evidences trends
- Stimulates thoughts relative to prevention
- Demonstrates management commitment and concern
- Identifies weaknesses in the safety program
- Justifies expenditures
- Serves to confirm or refute compensability
- Provides evidence for the subrogation of a claim
- Reduces future workers' compensation premiums

Figure 3.4

3.3 Hazard control

A three-tier hierarchy of controls — engineering, administrative and work practice, and personal protective equipment — is widely accepted as an intervention strategy for controlling workplace hazards. It's important to note that the best results are obtained when control methods are implemented in this order. But it's also important to note that some methods will not work, cannot work, or are not affordable. In this case, the next level of control should be attempted. The last resort, unfortunately, is often the employer's first and only attempt of hazard control because it appears to be a quick and easy solution.

3.3.1 Engineering controls

The preferred approach to prevent and control workplace injuries is to design the job or worksite so that the hazard is eliminated. Engineering controls involve physical changes to the workstation, equipment, facility, or any other relevant aspect of the work environment. Generic examples include eliminating toxic chemicals and substituting nontoxic chemicals (substitution is the solution), enclosing work processes or confining work operations, and installing general and local ventilation systems. Specific examples of engineering controls include using electrically adjustable beds as a substitute for manually adjustable beds, needleless systems to prevent needle sticks, forklifts to move product, or welding curtains.

One of the most prevalent, most expensive, and most elusive workplace injury categories is ergonomics, or the lack thereof. Mention of ergonomics is made here because evidence of inadequate control of ergonomic risk factors is present in nearly every company visited by the authors. The numbers of employers and employees alike who will contort their bodies to adapt to the concessionary configuration of $3,000 worth of technology on a $100 piece of furniture is truly amazing. Engineering control strategies for these hazards include actions such as

- Changing the way materials, parts, and products can be transported — for example, using mechanical-assist devices to relieve heavy load lifting and carrying, or using handles or slotted hand holes in packages requiring manual handling.
- Changing the process or product to reduce worker exposures to risk factors; examples include maintaining the fit of plastic molds to reduce the need for manual removal of flashing, or using easy-connect electrical terminals to reduce manual forces.
- Modifying containers and parts presentation, such as height-adjustable material bins.
- Changing workstation layout, which might include using height-adjustable workbenches or locating tools and materials within short reaching distances.
- Changing the way parts, tools, and materials are to be manipulated; examples include using fixtures (clamps, vise-grips, etc.) to hold work pieces to relieve the need for awkward hand and arm positions or suspending tools to reduce weight and allow easier access.
- Changing tool designs — for example, pistol handle grips for knives to reduce wrist-bending postures required by straight-handle knives,

or squeeze-grip-actuated screwdrivers to replace finger-trigger-actuated screwdrivers.

- Changes in materials and fasteners — for example, lighter-weight packaging materials to reduce lifting loads.
- Changing assembly access and sequence — for example, removing physical and visual obstructions when assembling components to reduce awkward postures or static exertions.

Ideally, occupational hazards are anticipated prior to their introduction into the workplace, when facility design, production lines, production rates, tools, equipment, and machinery choices and selections are determined. All too often, though, this is not the case. Whenever the authors have occasion to visit an insured who is contemplating the renovation of his current facility or the construction of another one, this point is driven home! In fact, the recommendation is made to include a safety engineer in the design process. The input of a safety engineer may seem counterproductive and expensive, but in the long run the elimination of a workplace hazard is well worth the investment in his or her services.

3.3.2 Administrative and work practice controls

Administrative controls are procedures that significantly limit daily exposure by control or manipulation of the work schedule or manner in which work is performed. Administrative controls do not eliminate or limit the hazard. Consequently, the controls must be consistently used and enforced. Examples include

- Controlling employees' exposure by scheduling production and tasks, or both, in ways that minimize exposure levels — the employer might schedule operations with the highest exposure potential during periods when the fewest employees are present.
- Reducing shift length or curtailing the amount of overtime.
- Scheduling more breaks to allow for rest and recovery.
- Rotating workers through several jobs with different physical demands to reduce stress on limbs and body regions.

Work practice controls alter the manner in which a task is performed. Some fundamental and easily implemented work practice controls include

- Changing existing work practices to follow proper procedures that minimize exposures while operating production and control equipment.
- Inspecting and maintaining process and control equipment on a regular basis.
- Implementing good housekeeping procedures.
- Providing good supervision.

- Mandating that eating, drinking, smoking, chewing tobacco or gum, and applying cosmetics in regulated areas be prohibited.

3.3.3 Personal protective equipment

Personal protective equipment (PPE) is specialized clothing or equipment worn by an employee for protection against a hazard. PPE typically is used when other engineering and administrative/work practice controls are not feasible or until other controls can be implemented. Traditionally, PPE serves as a supplement to minimize employee exposure, not as a primary source of control. Examples of PPE include, but are not limited to, rubber boots, gloves, face shields or masks, and eye protection. PPE must be accessible and provided in appropriate sizes at no cost to the employee. The employer also must ensure that protective equipment is properly used, cleaned, laundered, repaired or replaced as needed, or discarded.

The greatest challenge faced by employers who implement the use of personal protective equipment as a means of hazard control is supervision. Let's face it — wearing PPE is a hassle. It's another thing that employees must maintain and put on their bodies. Seeing, hearing, and feeling can be challenging enough without the added layer of protection. Quite often the hazard is not readily apparent to the employee, so the necessity for PPE does not seem important. The result is that many employees will not wear the PPE without direct supervision or constant reminders. Supervisors get annoyed and then must balance productivity (and harmony) with compliance. All too often, lax enforcement of the PPE policy undermines the program. The negative feelings regarding the use and inconvenience of PPE is transferred toward "safety" and not toward the actual hazard. Ultimately, this transference of negative feelings and attitudes can whittle away at the entire safety program until all preventive interventions are perceived as disruptive measures imposed by some safety regulation. This may seem dire, but the fact remains that PPE, which should be the last resort, often is the employer's first and only attempt at hazard control. If at all possible, eliminate the hazard.

3.3.4 Control evaluation

A follow-up evaluation is necessary to ensure that the controls reduced or eliminated the hazard and that new hazards were not subsequently introduced. The authors visited an insured (farm equipment and implement dealership) who had been consistently incurring eye injuries. Even though employees were wearing protective eye wear and following established safe work practices, the frequency of these injuries remained at a high level. Upon inquiry, it was determined that the type of safety glasses in use was not designed, and obviously not effective, for precluding the passage of particles around the sides of the glasses. Because of how the employees did their work, on their backs and sides, under the tractor or implement particles were entering the employees' eyes. Again, all hazard controls, whether engineering, administrative, or personal protective equipment, must be evaluated to ensure that they are effective and that new hazards are not subsequently introduced.

3.4 Performance assessment

3.4.1 Why do it?

Some question the utility of studying the summary of an organization's safety performance since it's retrospective and has little or nothing to contribute to the future. There is some merit to this point, but only in the definition that it is a retrospective analysis. Otherwise, loss history analysis is a bed of information for assessing what went wrong, where it went wrong, who and what was involved when it went wrong, how often it went wrong, and, we hope, why it went wrong. Synthesizing this data into a plan of action for controlling the opportunity for recurrence is prospective. Implementing that action plan is *proactivity*. Thus, understanding where you are is necessary information in the process of getting to where you want to go.

Investing vast amounts of time on data collection and analysis, and then becoming a "bean counter" who spouts injury facts is a common pitfall of some who do the assessment. Don't do it. It serves no purpose and often alienates those who could use the data, in its proper context, to preclude further injuries.

Systematically assessing the safety performance of your organization is comparable to taking its pulse — it's not complicated and it's nonintrusive. What's required is that a system be established. Organizations external to your company, such as the insurance industry and the government, already have systems in place. The task, therefore, is to simply create a compatible system internally or modify current procedures so that interface with these external sources of data can be accomplished with ease on a routine basis. Simply put, external data is available to assist in the assessment of your company's safety performance, but without similar data collected internal to your company, no comparison can be made.

Understanding where the "soft spots" or gaps exist in your safety program enables you to prioritize corrective action. Safety performance assessment identifies these deficient areas. Shoring up these gaps will preclude recurrence of those injuries. Fewer injuries translate into less money spent on workers' compensation expenses, which leads to a higher profit margin for the company. Yes, loss history analysis is retrospective, but what can be accomplished with the information gained from that look into the past can positively affect the cost of your workers' compensation insurance in the future.

3.4.2   Types of assessment

3.4.2.1   Trending

Data collection and analysis that is focused solely on the organization's safety record and then measured against the same parameters, but for a different time period, is known as *trending*. The objective is to identify patterns or cycles, if any, in the occurrence of accidents. Some may be relatively easy to delineate — which department sustains the most injuries, which individuals are injured most often, what body parts are typically affected, or if a production cycle corresponds with an increased frequency of accidents. Other patterns may be more difficult to discern — was new equipment training presented by a contractor deficient and contributory to a surge in accidents involving the equipment, is there an increase in substance abuse among your employees, or has a recent substitution of fluorescent light bulbs in the production area altered the illumination enough to challenge some employees' vision?

Collecting the data should be systematic, but analyzing it should be exploratory, inquisitive, and critical. Look at it! The answers to the questions of who, what, when, where, and why are usually buried in the facts and figures. Compare yourself to a detective who is seeking clues in an attempt to identify the reason for the accident.

3.4.2.1.1   Incident rate.   Incident rate is a common computation that allows an organization to check its safety record (as measured by injuries) against either itself over a period of time or other similar organizations. When used as an external measurement, it is considered benchmarking, which is discussed below. The "magic" of incident rates (IR) is that the formula, which determines the rate, relates the number of injuries to the number of employee-hours worked. Therefore, small companies and large companies can be compared. What results is a rate that can be compared to other time periods or among departments. The employer can measure accident trends by simply computing the IR over a period of time and plotting it; the graph will visually display "unsafe" periods, if any. From there, the astute employer can dig deeper to determine why a particular period of time had more accidents. That same employer can also compute and then compare the IR for each department. Obviously, this information can be used to focus company safety efforts in those areas that need it more than others.

The formula for computing an incident rate is depicted in Figure 3.5. The constant in the IR formula is the number 200,000, based on 100 employees working 40 hours per week for 50 weeks per year. The variables are the number of accidents (incidents) incurred by your company and the number of hours worked by your employees during the covered period.

3.4.2.1.2   Loss-run analysis.   Compiling the information pertinent to workers' compensation claims into a "user-friendly" format provides an

Incident Rate =

$$\frac{\text{\# of injuries and illnesses X 200,000}}{\text{Total hours worked by all employees during period covered}}$$

Figure 3.5

employer the opportunity to study the facts and consider any relationship that may emerge. Some businesses are able to administratively track this data because of their staffing or level of sophistication, but most are not. This data is available to all businesses, however, and can easily be obtained from your insurance carrier. Ask for it.

Typical loss runs depict the loss history of your company during its period of insurance coverage by the current provider. Previous insurance carriers retain the loss runs during your policy with them, so they too should be contacted and requested to provide a copy of that portion of your company's loss run if it was never received.

The data included in loss runs usually consists of the claimant's identification, a brief synopsis of the injury, a chronology of significant events (date of injury, first day off work, date of return to work), and a tally of the financial expenditures for compensation, medical attention, litigation, and rehabilitation. For those organizations that have multiple locations or accounting centers, codes can delineate specific departments in which the injuries originated.

It's easy to see what a great management tool a loss-run analysis is. If trends are present, they will emerge after only a few minutes of studying the data. Trends that can emerge from the data include

- Which department sustains the most injuries
- Body part most often affected by injury
- Time of year when injuries peak and ebb
- Particular employees who appear on the loss-run more than others
- Consistently excessive period of time between the date of injury and return to work
- Discrepancies between accounting of financial expenditures and that presented by the insurance carrier
- Correlation between an unacceptable loss experience and other less-than-desirable performance indicators in certain departments.

The more the data is analyzed, the greater its utility.

3.4.2.1.3   Accident investigations and near-miss reviews.   The usefulness of conducting accident investigations is thoroughly discussed in Chapter 8, but the point is made here to emphasize their criticality. The description of the events surrounding an accident is the closest you will ever come to being there when it occurred. Without this information, or without scrutinizing what information is captured during the investigation, the opportunity to reenact the scenario mentally is lost.

Accident investigations and near-miss reviews are strictly an internally focused analysis. However, if that analysis results in the suspicion that an underlying causative factor may be something external to the company (such as a piece of equipment or chemicals used in the process), then the focus should extend outside the company to determine if others are experiencing similar mishaps. Sources of information concerning the accident history of equipment or chemicals include the manufacturer, importer, or distributor. Several government agencies, such as OSHA, the National Highway Transportation Safety Board, or the Consumer Product Safety Commission, are also repositories of accident information. Check their data to determine if the accidents or near-misses occurring in your operation have any resemblance.

3.4.2.2   Benchmarking

Just as trending is an internally focused assessment of performance, benchmarking is an external focus. Quite often business owners are interested in the effectiveness of their safety programs in comparison with their competitors or even dissimilar businesses. This comparison can and is made in the insurance industry when premiums are established. Assessing your safety program — through the company's loss history — against other organizations on a national level is a wise evaluation. Primarily, two comparison methods — incident rate and experience modification — are utilized.

3.4.2.2.1   Incident rate.   As discussed above, your company IR is a measurement unit that can be compared to any other organization in a similar industry, without regard for the size of the organization. Assuming that the rate has been computed correctly, the next step is to determine the rates for other organizations. This has been accomplished through the system established in OSHA's record-keeping requirements and is readily available from the OSHA Internet web site or through the mail.

3.4.2.2.2   Experience modification rating.   The term "experience modification rating" is unique to the insurance industry. In essence, it is your company's safety reputation, a numerical rating that quickly describes your loss experience in comparison to similar organizations. The computation of the rating and its importance are discussed in greater detail in Chapter 5.

This figure is adjusted annually, dependent upon the losses incurred by your company. The more losses incurred, the higher the figure. Companies with high modification rates have sustained more losses than those with low modification rates. Your experience modification rating can be used as a benchmark since it is prepared by the insurance industry, with national data, and it compares your company's performance against similar companies.

3.5 Summary

To reduce the incidence of occupational injury and subsequent workers' compensation claims, employers must develop an awareness of the hazards that can exist in the workplace, identify those hazards that do exist in their workplace, and then develop methods of controlling or eliminating those hazards. The process of doing so should be systematic and should include employees, supervisors, and outside sources. Employers must understand, too, that hazards evolve so the identification and control process is just that, a process, similar to the one displayed in Figure 3.6. The process should be continuous, commensurate with the continuity of the production process. It may appear to be an overwhelming task. But with the assistance of those who face the hazards daily, their supervisors, and the safety professionals external to the organization, it can be accomplished and then become a matter of routine. Occupational injuries will then decline.

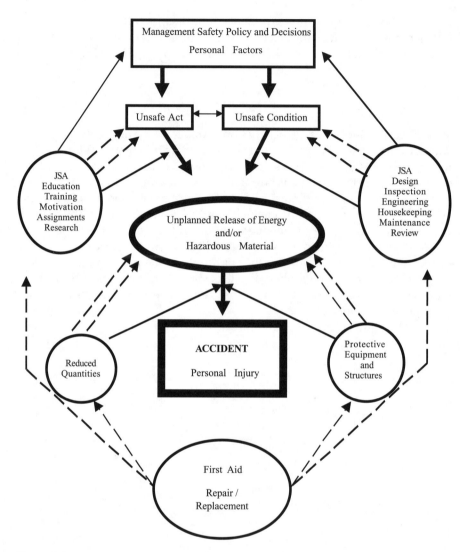

Figure 3.6

chapter four

Safety Internet resources

Contents

4.1   Overview of the Internet

Much has been said regarding the information explosion, computer technology, and the Internet. It is a rapidly expanding means of information management that the business world is incorporating into profitable practice. New applications are devised daily, some of which directly affect occupational safety and health and the minimization of workplace injuries.

It is estimated that thousands of individuals, agencies, companies, and institutions are joining the Internet each day, both to share and to acquire information. This unprecedented connectivity makes the introduction of the telegraph, telephone, and television pale in comparison. Today, almost every citizen in this country can now easily access information that was virtually unobtainable only a few years ago.

The origin of the Internet can be traced to a few organizations and institutions, including the national defense establishment, universities, computer hobbyists, and even some commercial entities. Their electronic information resources remained an internal source of data until the advent and commercial proliferation of fiber-optic cabling, individual modems, and network servers. This affordable, user-friendly technology then provided non-affiliated individuals and organizations the means of accessing these data files. The result has been the addition of another dimension to the information explosion of the late twentieth century and a sample of the quantum changes that technology will introduce during the twenty-first century.

4.2   Accessing the web or "going online"

Several methods of connecting to the Internet exist. You may do so from your own home, or, if your business has already incorporated computer technology into traditional business tasks, you probably have Internet access there. The public library is another source of Internet access. To have Internet access from home, in addition to having a computer and modem, you will need to select an Internet Service Provider (ISP) and determine how many hours of access you estimate using on a monthly basis. You can start this process by asking a friend who has already been through the process or by looking in the Yellow Pages for Internet Service Providers. It's interesting to note that during the previous 24 months, the author has changed ISPs four times — three as the result of corporate buyouts and mergers and one due to local telephone access discounts. The Internet is in its infancy but it's growing rapidly; so, too, are the logistics involved with participating in the online dialog of information exchange.

4.3   Advantages provided by the Internet

Before focusing on the specific application of Internet resources to the reduction of occupational injuries, an overview of this electronic information retrieval system's capabilities is warranted. As previously mentioned, the Internet is conveniently accessed, but it is also instantly accessible 24 hours a day. Sites can be visited at lunch, break time, or during the early hours of the morning. The Internet never closes.

The Internet offers a variety of domains from which to select data. Typical domains include government agencies, educational institutions, private organizations, and commercial sites. Foreign sites are also online. Some sites are very credible; others are more opinion than fact. Some are cluttered with advertisements. The web site address will contain the three letters that identify its domain — gov, edu, org, com. What you are after will quite often dictate which domain to search.

Searching the web can be comparable to looking for a needle in a haystack unless you use a search engine — a web site that searches the Web for you. There are several search engines available for use. If one does not focus narrowly enough for your needs or if it does not find the quality of information you desire, then you can try another search engine and search again. Searches usually take much less than a minute. Commonly used search engines include Yahoo!, Excite, and AltaVista. You'll know that you need to refine your search when you get thousands of "hits" (data returns) from your inquiry. It takes practice and experience. If you have the Internet address of an associate company or a regulatory agency, you can enter the address in your web browser (such as Netscape or Explorer) and go directly to the site.

One of the best assets of the Web is that most quality sites link to additional information. One might compare this to a reference section or bibliography, but it is unique to the Web because it is a listing of other web sites that contain related data. If a particular site does not prove beneficial, its links might provide the information you want. All the researcher needs to do is simply "click" on the link and that information source will appear on the computer screen. It is amazing how fast the time passes as one clicks from web site to web site, rapidly sorting information. Quite often, items of interest other than the intended data will divert your attention, and this is where the researcher must exercise discipline and remain on task.

Ironically, the greatest advantage of the Internet is also its greatest disadvantage. Information overload can frustrate and confuse.

4.4   The WIIFM formula

You might be asking, "What's in it for me?" or "How can the Internet increase the productivity of my organization by improving my safety management program?" These are valid and appropriate questions.

There are numerous ways that the Internet can be useful to business. As more individuals and organizations incorporate the Internet into their daily approach to safety management, more and better methods will evolve. Listed below are five considerations.

- Regulatory data is prolific and difficult to locate in hard copy. Not all libraries maintain the Code of Federal Regulations, the National Electrical Code, and numerous other laws that affect the workplace, but these documents can be found on the Internet.
- Laws, rules, and standards change. One can read the Federal Register daily on the Web or visit the web sites of affected organizations and read their editorials.
- Quality safety programs are available on several web sites, free for downloading and subsequent adaptation to your specific business operation.
- Training programs are also offered, most for free, and range anywhere from toolbox safety talks to regulatory compliance presentations. One need not spend time composing lesson plans — they are available online.
- Electronic information forums exist online and provide the opportunity for participants to monitor the challenges, solutions, and activities that other members (who may have common interests, pursuits, or responsibilities) have willingly posted.

4.5   Information management

There are numerous sites of particular interest to those interested in occupational health and safety. A selection of these are described below.

4.5.1   Government agencies (.gov)

If you're looking for regulatory data, statistical data, current events, or the status of impending change, government sites are unparalleled. Their content is typically current, vast, and relatively free of bias. However, the government is politically flavored and influenced, so these sites may occasionally contain such editorializing.

- Occupational Safety and Health Administration (www.osha.gov). This is the oracle Internet site for safety and health. It contains information concerning all aspects of workplace hazards (chemical, ergonomic, physical, environmental, and even violence) and suggested means of abating the hazards. Inspection results are posted on this site for public access; if OSHA has visited you, the results of that inspection are available for scrutiny by customers, competitors, insurance companies, and anyone else who has access to the Internet. Agency interpretation of regulations, their instructions to inspectors, and their response to letters of inquiry add depth to this site. Links to additional useful government sites are provided.
- Department of Transportation (www.dot.gov). For those organizations that deal with transportation, this government web site should be a regular pit stop. Not only does it address the myriad transportation regulatory requirements, but it also includes information relating to driving safety. For example, under the category of Frequently Asked Questions (FAQ), topics are addressed such as how to order safety-related materials and research conducted concerning driving and the use of cellular telephones.

Perhaps its best link is to the Office of Motor Carrier and Highway Safety, a subset of the DOT, which lists safety programs, facts and figures, rules and regulations, and related links.

- Environmental Protection Agency (www.epa.gov). The EPA affects all companies, regardless of size or industry. This web site includes the Small Business category where topics ranging from regulations to saving money are discussed. Incidentally, of the government sites, this web page is one of the more aesthetically pleasing visual displays.
- Federal Emergency Management Agency (www.fema.gov). Emergency response plans or emergency action plans are necessary for most businesses. Depending upon the number of employees or the complexity of operations, these plans must be written. The FEMA web site

is a tremendous source of information in this regard. In fact, an emergency management guide for business and industry is available, free of charge. This guide is a step-by-step approach to emergency planning, response, and recovery for companies of all sizes and is applicable to tornadoes, fires, earthquakes, blizzards, floods, explosions, and civil unrest. This site also links to the Interactive Hazard Map (part of Project Impact, a joint program with FEMA and ESRI) that will plot the recorded locations and dates for significant natural disasters (floods, earthquakes, windstorms, hailstorms, tornadoes, and hurricanes) within any particular region based on zip code, city and state, or U.S. Congressional District.

- Centers for Disease Control and Prevention (www.cdc.gov). Although at first the CDC site might seem a bit off-track as a source of information for the typical business owner, the organization conducts significant research and data collection that contributes to occupational safety and health. The site provides both in-depth description of research and one-page fact sheets. The fact sheets can serve as great discussion topics during safety meetings or can be reproduced and included as paycheck stuffers. Some applicable topics include back belts, carbon monoxide poisoning, health and safety manuals, hepatitis, lyme disease, motor vehicle-related injuries, occupational health, and tuberculosis.
- National Institute for Occupational Safety and Health (www.cdc.gov/niosh). NIOSH is a subset of the Centers for Disease Control and Prevention. Its specialty is research, training, and education in support of occupational safety and health. One could say that in the realm of occupational safety and health, NIOSH is the brains while OSHA is the brawn. The two organizations are distinctly separate, with NIOSH operating in a much less political environment than OSHA. The information available through this web site is nearly infinite and addresses almost every category of occupational endeavor. Quick reference topics include agricultural safety and health, chemical safety, material safety data sheets, construction safety and health, ergonomics and musculoskeletal disorders, healthcare workers, indoor air quality, industrial hygiene, noise and hearing protection, respirator information, stress, and workplace violence.
- Department of Labor (www.dol.gov). The Department of Labor's site could easily occupy your entire day. Much of the information available from the DOL may seem removed from the interests of occupational safety and health, but reconsider. How many employers have been challenged with the detrimental effects of employee substance abuse and its secondary effects on productivity? Do you have a drug and alcohol policy and program in your organization? How many employers are encountering the management challenges of the Welfare-to-Work Program? Have you ever wanted to read the applicable labor/management law (referred to as elaws on this site) that specifies

requirements pertinent to the above-mentioned situations? All this and more is one keystroke away. Visit this site to obtain the answers. You can even access the Substance Abuse Information Database to gain an appreciation of the seriousness of this pervasive undermining of our national industrial and commercial productivity. Information contained on this site may even assist your efforts to reduce employee turnover.

- Bureau of Labor Statistics (http://stats.bls.gov). In order to compare your company's loss history to that of other similar organizations, you must have access to national data. This site is that location and it is accessible 24 hours a day. In addition, this site can enhance the competitive status of your business by providing insightful information relative to several other facets of business activities.

- Department of Justice (www.usdoj.gov/crt/ada). This Internet address will take you to the Americans with Disabilities Act (ADA) home page, a source of valuable information that most, if not all, employers have needed or will need. Employees may be hired with a disability or they may incur a disability after employment. Either way, the ADA dictates the rights and responsibilities of both employers and employees. How does the ADA apply to safety and health in the workplace or workers' compensation? A better question might be if your company's modified duty/return-to-work policy and program comply with the Americans with Disabilities Act. Visit this site for answers. Provided is a toll-free phone number for those who wish to speak directly to a DOJ representative; a subsite that explains enforcement procedures under the law; another subsite that provides technical assistance and even some materials; and finally, a subsite that describes the ADA Mediation Program. Too much information — not enough time! But at least you know where to go to get the answers.

- State of California OSHA (www.dir.ca.gov/occupational_safety.html). The largest, most active, and sometimes labeled the benchmark state OSHA program, California OSHA is second only to the federal OSHA in terms of information and resources. A visit to its web site is highly recommended, especially if a portion of your work force is Hispanic, because CALOSHA is virtually bilingual in all of its offerings. Incidentally, for those businesses that are domiciled in other states, those states will have their own OSHA web site, provided that they have an authorized state plan. Otherwise, they are regulated solely by the federal OSHA. If you want to know which states have authorized plans, simply visit the federal OSHA web site and click on "state plans."

- State of Kentucky OSHA (www.state.ky.us/agencies/labor/). Only one of 40+ others, the Kentucky site is listed as an example of state sites. Realize that each one is different, with strengths in certain areas. Occasionally visit a different one for comparison. The Kentucky OSHA site is filled with choices — an overview of the organization, a section that contains new regulations, proposals, meetings, and notices; a section that explains complaints; another section that describes the enforcement aspect of OSHA; a section devoted solely to training; and a section that addresses the technical assistance available from the agency. Of particular interest is the "post-inspection guide," an instruction pamphlet developed to assist business owners in their actions after an OSHA visit. Available in the enforcement section, this guide can be downloaded and printed at your location — after, during, or even before OSHA ever visits. Forewarned is forearmed. A listing and brief description of the training sessions offered by Kentucky OSHA personnel are included in the training section. Topics include back care, accident investigations, fire safety, lockout/tagout, office safety, and much more. Last, the technical assistance section provides guidance in the correction of hazardous conditions.

4.5.2   Educational institutions (.edu)

Colleges and universities also maintain web sites, and some of them are wonderful caches of data, research results, and opinion. As long as you recognize opinion when you read it and place that source in its appropriate category (fact vs. fiction), these sites can add another dimension to one's inquiry.

Usually an educational institution will specialize in a particular area of interest. For example, some schools concentrate on ergonomics while others research fire science, chemical safety, or loss prevention. Whatever their motivation or objective, these institutional web sites are gems of information and contain links for further inquiry.

- Oklahoma State University (www.fireprograms.okstate.edu). This school has one of the premier academic programs with regard to fire science and works closely with the International Fire Service Training Association. The links provided at this site are wonderful stepping stones for additional search of related topics such as fire codes, fire extinguisher training, or storage of flammable materials.
- Cornell University (ergo.human.cornell.edu/). This is an academic one-stop shop for the latest in ergonomics/human factors research and information. Software programs designed to assist in the evaluation and remediation of worksite ergonomic hazards are offered for free downloading. Another advantage of academic sites is the capability to correspond personally with professors or students who are involved with the research. Perhaps your business operation contains a unique ergonomic challenge and you would be interested in allowing an academic research team to evaluate, study, and recommend a state of the art approach to the hazard.
- University of Minnesota (www.bae.umn.edu/~fs). Although this university is geographically on the periphery of the country, the

information available from its web site is applicable to all agricultural operations in the nation. Topics include farm safety, rural issues, and research and outreach programs. As with all quality web sites, it has additional links to related sites.

- University of Vermont (siri.uvm.edu/msds/). Probably without exception, all businesses and companies maintain one or more hazardous substances onsite. Each of these substances has a material safety data sheet (MSDS) that describes what harmful effects the substance can have on humans, what the symptoms of exposure are, and, most importantly, what to do in the event of exposure. An MSDS is supposed to be provided to the consumer by the manufacturer, importer, or supplier. When the MSDS is not provided or needs to be replaced, you can use the University of Vermont web site to get the replacement. The University of Vermont has one of the best MSDS retrieval engines on the Web.

- University of Virginia (keats.admin.virginia.edu). Most of what is available on this web site is designed for use by university faculty, staff, and employees in the accomplishment of routine duties, but it has application to all worksites. Topics include asbestos, biological hazards, chemical hazards, ergonomic hazards, and even the broader category of industrial hygiene. The appeal of this site is its readability and understandability — a great place to go for fast facts.

4.5.3   Professional organizations (.org)

Professional organizations often provide cutting-edge information and application pertinent to their specific interest. Some may very well be described as a cornucopia of electronic safety management resources.

Many of us can readily identify with a professional organization and its services (e.g., American Red Cross) but there are others that also affect the management of safety and health for American workers and which maintain web sites. Several of them have existed longer than the governmental agency that now regulates the industry. These sites are valuable sources of information and for the most part are freely accessed.

- National Fire Protection Association (www.nfpa.org). Probably few people know that the National Electrical Codes and Life Safety Codes are generated through the administrative efforts of the NFPA. Every structure involved in occupational activities must meet certain construction standards and the NFPA is a proponent. How can average employers ascertain their compliance or remain current regarding pending changes? Or, how can average employers improve their electrical safety training program? The answer is by becoming *above* average — simply access the National Fire Protection Association web site and read. This site is probably the best source for data pertinent to fire safety. Sparky the Fire Dog will host your visit and will welcome your inquiry.

- National Safety Council (www.nsc.org). This organization is one of the forerunners of safety information for American homes and job sites. Much of what it offers in terms of programs and information is designed for ease of understanding and immediate application. Members are entitled to even more options, but much is available to non-members. Information ranges from graphs and statistics to short essays to lesson plans and even sample safety programs. For decades the National Safety Council has functioned as a clearinghouse for safety information. This organization should be one of the initial sites visited by a novice in safety. It is a premier location.
- A variety of organizations with Internet addresses are listed below. Even if the name does not appear to have any relation to your concerns, visit it to check its listing of links.
  - American Society of Safety Engineers (www.asse.org)
  - American Industrial Hygiene Association (www.aiha.org)
  - Local Government Environmental Assistance Network (www.lgean.org)
  - Human Factors Ergonomic Society (www.hfes.org)
  - American Red Cross (www.redcross.org)
  - Institute of Hazardous Materials Management (www.ihmm.org)

4.5.4   Commercial entities (.com)

We are hard-pressed not to hear the term "dot com" on a daily basis — CocaCola.com, Ford.com, ABC.com. We even see it on advertisements next to the traditional telephone numbers. As mentioned earlier, the letters at the end of a web site address identifies the general purpose of the site or those that maintain it. Commercial sites end in .com and can be either very informative and beneficial or a waste of time. Once you find one that is credible, mark its location and save it for future reference.

The sites described below are credible and may very well contain solutions for some of your occupational safety and health program challenges. Check them out.

- www.safetyinfo.com — If this site were a restaurant, it would be rated with five stars. It is sponsored by several companies that profit from the safety industry (apparel, devices, consulting, etc.) but the information available is provided with no strings attached. It provides free samples of safety programs (such as bloodborne pathogens exposure control, fall protection, confined space entry, hazard communications, emergency action plans, office safety, etc.). There is no need to wait for an OSHA inspector or a private consultant to provide it. It also has lesson plans, with graphics, that can be downloaded to your computer and then printed. Recent developments in the world of occupational safety are also posted on this site. If you look nowhere else, at least check this site out. Some safety professionals call it their favorite.

- www.ohsonline.com — This site is the online version of *Occupational Health and Safety Magazine* and is an easy method of remaining current in safety trends. Article topics cover the entire spectrum of occupational safety and health.
- www.alimed.com — Need some material solutions (furniture, personal protective devices, workstation modifications) to your ergonomic challenges? Not only is this site an electronic catalog, it is also a repository of information pertinent to ergonomics. There is no need to wait for the salesman; the products are accessible online.
- www.ambest.com/safety/ — This site is another online catalog, but it is more inclusive than Alimed's site. It also contains some quality articles addressing occupational safety and health issues, which add credibility to this commercialsite.
- www.ehsfreeware.com — Not only is information free on the Internet, but software can also be freely downloaded from some locations, and this site is one such location. You may not get the newest and greatest, but you will have a selection of programs to assist in the management of both your safety and environmental programs. Use of these programs will help you decide whether additional, expensive commercial software is needed.
- www.free-training.com — This site allows the user to conduct select occupational safety and health training online. Although computer-based training should not be considered a sole source of training, it definitely can enhance any training program. If for no other reason, visit this site to compare its presentation methods and graphics to those that your company is currently using. It's always beneficial to see how other trainers approach the same subjects that you periodically present to your employees. Their approach may be more innovative and better retained by participants.
- www.ergonext.com — If you need to get smart fast on ergonomics, then go to this site first. The data, resources, links, and format of this site make it a tremendous tool for a safety manager. You'll be impressed with the contents and its presentation.

4.5.5 Listservs

This category is unique because it typically is a subset of another domain. Many describe it as electronic interactive bulletin boards. The advantage of subscribing is obvious — worldwide input. The challenges faced by your company may have already been addressed somewhere else. All you need to do is join the list, post your question, and await responses. Solutions may be minutes away.

Remember the old saying, "The only dumb question is the question that is not asked"? Ask your question through cyberspace and you cannot be embarrassed. Two of the better listservs are described below.

- Duke University (occ-env-med.mc.duke.edu/oem). This particular site, known as the occupational and environmental medicine listserv, is one of many offered by Duke University. It functions as an electronic bulletin board for the exchange of information pertinent to the pursuit of occupational medicine. Although this may seem too scientific or technical, it contains "need-to-know" information for the layperson, also. Questions are posted and then answered by list members (approximately 2,500), usually the same day. There is no charge for membership, but you do need an e-mail address. On a daily basis there are usually 10–20 questions, answers, statements, or announcements sent to the list, which are then automatically sent to your e-mail address. Topics have included carbon monoxide exposures, needle sticks, solid-waste handling, and Lyme disease vaccinations.
- University of Vermont (www.esf.uvm.edu). Known as the Safety List, this 3,500-member forum concentrates on occupational safety and health issues. For the safety professional, monitoring this list is as beneficial as all other research activities. Message traffic on this site is limited to 50 postings daily. Most days are full. An archive is available for the 10 years of postings. This listserv can keep you in touch with the reality of safety management across the country and even across the border.

4.6 Conclusion

It should be noted that the mention of specific web sites, addresses, and content is a "risk" for the author of traditional, printed material because books are static sources and "dated" as soon as they are published, while the Internet is dynamic and always being updated. Consequently, the reader should be aware that the above sites may change, or even no longer exist, by the time they read this text. The point here, however, is that a new source of information is available.

Our development of information management capabilities has consistently remained a step behind due to the unprecedented, rapid introduction of newer technology. It's critical that we monitor these developments, if only from the sidelines, so that we can identify that application which will leverage our safety program. To date, our imaginations have been the only limiting factor. Expand your horizons and include the Internet in your safety program. Get online or get left behind.

chapter five

Understanding and influencing workers' compensation premiums

5.1   Introduction

For company owners and managers entrusted with the responsibility of decreasing workers' compensation costs, it is essential that they first understand the manner by which workers' compensation premiums are calculated. It is only then that effective measures can be planned and implemented to achieve that goal. Similarly, individuals who are charged with the administration and management of company safety efforts must also be cognizant of how workers' compensation premiums are calculated, as the impact that specific safety initiatives may have on future premiums can be a significant justification for their implementation.

The intent of this chapter is to provide an analysis of how workers' compensation premiums are determined, followed by practical advice on the various ways by which premiums can be influenced.

5.2   Calculating premiums

In brief, workers' compensation premiums are calculated in a three-stage process:

1. The manual premiumis determined.
2. Experience rating adjusts the manual premium.
3. Applying miscellaneous credits and debits.

Each of these procedures is described in greater detail below.

5.2.1   Determining the manual premium

There are a variety of different categories of employees within a single company, based upon the type of work the employees perform. Within the context of workers' compensation, the categories, which attempt to denote the type of work performed by individuals, are called "classifications." These classifications are the fundamental building blocks upon which the entire rating system is based. The most widely used classification system for workers' compensation was developed and is administered by the National Council on Compensation Insurance (NCCI), which currently includes over 600 different job classifications. Although no single classification describes every individual's job duties exactly, employees are categorized into the classification that best describes the work they perform.

Workers' compensation classifications are denoted by four-digit numbers, but it is important to note that these codes differ from Standard Industrial Codes (SIC). Whereas workers' compensation classification codes describe jobs performed by the employees of a company, SIC codes describe the operation of the company as a whole. Furthermore, a company normally has only one SIC code but generally has employees in several workers' compensation classifications.

To illustrate the use of classification codes, consider a small lumber and building materials dealer. Generally, such a business has employees who fall into one of at least three classifications — clerical, building materials store, and building materials yard employees. The individuals classified as clerical employees include administrators, payroll personnel, purchasing agents, and other employees who perform work in an office environment. The individuals classified as building materials store employees work inside the retail area stocking shelves, assisting customers, and serving as checkout clerks. Last, the individuals classified as building materials yard employees work in the lumberyard, operating forklifts and loading/unloading trucks. These "yard" employees also include those who deliver merchandise to customers. As evidenced by the latter classification description, employees with similar risk exposures are frequently grouped within a single classification. Within that classification, delivery drivers are grouped with employees who actually work in the lumberyard area, as both are exposed to many of the same hazards. Furthermore, in many lumberyards, the same employees who work in the yard area also make deliveries.

For each classification, individual insurance carriers establish rates. Hence, the amount charged for any given classification varies from one insurance carrier to another. These rates are not arbitrary but are based on the actuarially derived risk associated with each classification. Therefore, classifications with a very low risk, such as clerical employees, are ascribed a low rate. Conversely, classifications with a high risk, such as underground coal miners, are assigned a significantly higher rate.

Rates are expressed in the amount that the company must pay for workers' compensation coverage per $100 in payroll. Using the above classifications as an example, the rate for the clerical employees may be $0.28 per $100 of payroll, whereas the rate for underground coal miners may be $30 per $100 of payroll.

Although the rates charged for each classification are based primarily on the risk associated with the work performed by the respective employees, these rates also include anticipated administrative expenses and overhead costs that the insurance carrier must incur to remain in business. These administrative expenses and overhead expenses are generally a monetary constant. As such, the same amount is added to the rate for each classification.

Inasmuch as the rates charged by insurance carriers are based on quantitative data such as risk, administrative costs, and overhead expenses, they are also affected by less scientific influences. One such influence is the open market economy. To a large degree, rates are affected by competition. When competing insurance carriers lower rates to attract business, other carriers are forced to follow suit to retain their desired market share. This competition inevitably drives rates lower than what is necessary for the individual insurance carriers to be profitable. Therefore, a cyclic effect results, in which the market repetitively softens and stiffens. During a soft market, rates are at their lowest and employers reap the benefits.

Upon applying for workers' compensation insurance coverage, the employer is asked to estimate the company's payroll for the upcoming policy year. This estimate must separate the anticipated payroll into each of the aforementioned classifications. To derive the manual premium, the total estimated payroll for each classification is multiplied by the corresponding classification rate. The employer's manual premium is the sum of the individual classification premiums. Figure 5.1 illustrates a worksheet that underwriters might use to calculate manual premiums.

Because the insurance carrier bases the workers' compensation premium on an estimation of anticipated payroll, an individual representing the insurance carrier must audit the company's actual payroll after the policy period has ended. This is done to determine how close the estimated payroll in each

Premium Calculation Worksheet			
Company: XYZ Lumber and Building Supplies, Inc.			
Address: P.O. Box 4399, Awatha, Nevada 57104-4399			
Contact Person: Mr. Robert Swafford		**Telephone Number:** (630) 555-3763	
Classification	**Payroll**	**Rate (Per $100)**	**Premium**
Clerical	$100,000	$0.28	$280
Inside Sales Clerk	$250,000	$2.08	$5,200
Lumber Yard and Drivers	$250,000	$4.34	$10,850
		Manual Premium	$16,330

Figure 5.1

classification was to the actual payroll. If the payroll was overestimated, the insured will receive a refund to compensate for overpayment. Conversely, an underestimated payroll will result in an additional bill from the insurance carrier.

5.2.2   Experience rating

Whereas the variance in rates for different classifications reflects the insurance carrier's attempt to anticipate the future cost of claims for each type of work performed, the system of classifying employees and charging different rates for each classification does not itself take into consideration the reality that within an industry some companies are more successful than their counterparts at preventing injuries and controlling the cost of claims. With this in mind, the manual premium is adjusted to reflect the anticipated losses of individual companies. This process is called *experience rating*, a process which attempts to predict the future losses of a company by mathematically comparing the company's actual incurred losses with the expected losses for each classification. The logic behind this process is that the past loss history of a company is the best available indicator of its future loss history.

Despite the tremendous impact that experience rating can have upon the net premium (the amount ultimately paid by the insured), many employers lack an understanding of how premiums are adjusted using the experience rating system. Not all companies are eligible to be experience rated. Each state has its own rules which generally include a minimum premium threshold and a specified minimum duration during which losses have been reported to the rating organization, such as a manual premium of at least $5000 and a reported loss history of at least three years.

An *experience modification factor* is the cornerstone of the experience-rating process. It is commonly called an "experience mod" or simply "mod." In short, it quantifies an individual company's past workers' compensation loss history through a complex process, which compares the company's actual incurred losses with the expected losses for each classification.

NCCI is the most widely used rating organization for workers' compensation and is responsible for calculating the experience modification factors of individual insureds for many insurance companies. As indicated previously, this organization also developed and administers the classification system used by many insurance carriers.

The frequency and severity of incurred claims affect the experience mod. As noted elsewhere in this text, frequency refers to the number of claims within a given period. Severity refers to the incurred cost of claims, both individually and collectively. These characteristics of a company's loss history are compared with the cumulative data submitted to the rating organization from all other companies doing business in that state that have like-classified employees.

Although both the frequency and severity of incurred losses are used in the calculation of an experience mod, it is notable that the frequency of losses has a greater influence. The reason for this is twofold. First, even losses with a low severity (little cost) involve administrative costs. These costs are not reflected in the reported incurred costs. Hence, a company that submits numerous workers' compensation claims, each incurring relatively small medical costs, still causes the insurance carrier to incur administrative expenses. Second, a company that has high frequency of losses, without regard to the severity of individual losses, indicates a company that does not have an effective means of preventing losses. To illustrate, compare two identical companies. One company reports 50 workers' compensation claims in a policy year, each claim incurring $1,000. In the same time frame, the other company reports one workers' compensation claim, which incurs $50,000. Although both companies incur a total loss of $50,000, the former is a less desirable risk since any of the 50 losses could have been more severe. Furthermore, the company that reported 50 workers' compensation claims is presumed to have a less effective mechanism for the prevention of injuries than does the company that submitted only one claim.

The experience mod is based upon the past losses incurred within a given period, generally three years. By using several years of loss history, the experience mod is not overly influenced by frequent or severe losses, which were confined to a short period. It does not include the losses incurred during the current policy year because that year is incomplete when the experience mod is being calculated for the upcoming policy year. As such, if a three-year period is reviewed to calculate the premium for the 2001 policy year, the loss history used to calculate that experience mod would be from the 1997, 1998, and 1999 policy years, not the loss history from the 2000 policy year.

Each year the rating organization issues a new experience mod, based upon the updated loss history of the company. This experience mod is generally issued prior to the policy renewal date, so it can be used to calculate the premium for the upcoming policy period. Notification of the experience mod is often sent by the rating organization directly to the employer. The employer then forwards that information to its insurance agent or directly to its workers' compensation carrier.

If a company has a loss history that mirrors the industry average, it is assigned an experience mod of 1.00. An experience mod of 1.00 has absolutely no impact on premiums. If a company has a loss history worse than the industry average, the experience mod will be greater than 1.00. This is called a debit experience mod. Conversely, if a company has a loss history better than the industry average, the experience mod will be less than 1.00. This is called a credit experience mod.

A premium that has been experience-rated is called a modified premium. In simple terms, a modified premium is a product of the manual premium and the experience mod. Hence, for a company with a manual premium of $100,000 and an experience mod of 1.37, the modified premium is $137,000. Conversely, if the same company had an experience mod of 0.75, the modified premium would be $75,000.

Although the intent of an experience rating is to gauge more accurately the future losses of individual companies based upon their own history of losses, it also serves as an economic incentive for employers and a tool for safety managers. As an economic incentive, the experience rating process rewards employers for effectively focusing on the prevention of injures and the containment of costs following an injury. As a tool for safety managers, the experience modification factor translates the value of safety into a quantitative instrument that can be used when making presentations or requisitions to administrators who seemingly focus purely on the bottom line. To this end, the experience modification factor can be used both to demonstrate the need for accident prevention efforts and to demonstrate the effectiveness of previously implemented efforts.

5.2.3   Calculating the experience modification factor

To avoid glossing over the actual computation of experience modification factors by calling it a complex actuarial task, this section provides a thorough understanding of the process. Without this knowledge, the employer is at the mercy of the rating organization to enter the proper information.

As indicated previously, the computation of an experience mod is the actuarial comparison of expected losses to actual incurred losses. To determine the experience mod for the upcoming policy period, data from the three preceding years are used, excluding the current policy year. The formula for computing the experience mod is the sum of the actual primary losses, the stabilizing value, and the actual ratable excess, divided by the sum of the expected primary losses, the stabilizing value, and the expected ratable excess. Figure 5.2 provides a mathematical illustration of this formula. As depicted in the figure, the computation of an experience mod is really nothing more than a simple division equation. Although oversimplified, the equation is basically the actual losses incurred by the employer divided by the expected losses during that period.

Split rating is a process used in computing an experience mod to give greater weight to accident frequency than to accident severity. It separates the incurred amount of individual losses into two categories, primary losses and excess losses. Primary losses include all incurred expenses of individual claims up to a certain amount (for example, $5000) depending upon the rating organization. If the entire incurred loss is less than $5000, the entire incurred amount of that particular claim is considered to fall into the primary loss category. If an individual claim incurred a loss greater than $5000, the balance is called the excess loss. However, each state has a maximum individual loss amount that can be used in determining experience modification factors. For that reason, only the amount of an individual claim that is less than that limit is used. For the purpose of calculating the actual incurred excess loss of any individual claim, any amount over the state limit is disregarded. The amount of each claim that exceeds the primary loss but does not exceed the state limit is called the *ratable excess loss*.

$$\frac{\text{Actual Primary Losses} + \text{Stabilizing Value} + \text{Actual Ratable Excess Losses}_w}{\text{Expected Primary Losses} + \text{Stabilizing Value} + \text{Expected Excess Losses}_w}$$

Figure 5.2

The *actual primary losses* for use in the experience mod formula (Figure 5.2) are the sum of all primary losses for the three-year period being reviewed.

To determine the *weighted actual ratable excess losses* (denoted in Figure 5.2 as Actual Ratable Excess Losses$_w$), the sum of all ratable excess losses for the three-year period under review is calculated. That sum is then adjusted by multiplying it with a credibility factor determined by the rating organization and based on the size of the exposure (payroll). This credibility factor reflects the confidence placed in past losses to more accurately predict losses in the upcoming policy year. Due to the law of large numbers, greater confidence is placed in the use of past losses of companies with larger exposures (payrolls) to accurately forecast future losses. Hence, the credibility factor increases in proportion to the premium size.

The *expected primary losses* for use in the experience mod formula (Figure 5.2) are computed by using figures from the rating organization. They are calculated by multiplying the expected loss rate per $100 in premium by the payroll in each classification. This product is then multiplied by $100 to produce the expected losses in each category. Just as was done with the actual incurred losses, these totals are separated into primary losses and excess losses. To determine the amount that is expected to constitute the primary loss for each classification, the expected loss must be multiplied by a factor called a D-ratio, an actuarially derived figure that is based on the amount of losses that have been primary for all companies submitting data for that classification over a period of years. The product of the D-ratio and the expected losses for each classification is the expected primary losses. The total of these primary losses from each classification is then determined for the three-year period being reviewed. Note that the expected loss rates are not the same as the rates charged by individual insurance carriers, but instead they are calculated by the rating organization or state workers' compensation entity as the average (mean) of the losses incurred in the respective classification by all of the companies in the state.

The *weighted expected excess losses* (denoted in Figure 5.2 as Expected Excess Losses$_w$) is calculated in the same manner as the weighted actual ratable excess, but the data from the expected losses is used. Hence, the sum of all expected excess losses for the three-year period being reviewed is calculated. That sum is then adjusted by multiplying it with the same credibility factor as was used to calculate the weighted actual ratable excess losses.

The *stabilizing value* is calculated by multiplying the expected excess losses with the complement of the credibility factor and then adding a tabular ballast value. The expected losses used to calculate the stabilizing value is not the weighted value of the expected losses, as used above, but is the unweighted expected losses value. Instead of multiplying the expected excess losses by the credibility factor, in this part of the formula the expected excess losses are multiplied by the complement of the credibility factor. Hence, if the credibility factor is 0.25, the expected excess losses are multiplied by .75 (1.00 − 0.25). A tabular ballast value is then added. This tabular ballast value is merely a statistical weighting factor, which is added to both the numerator and denominator to prevent excessive fluctuations in the quotient.

With each of the five elements of the formula used to calculate a known experience modification factor, anyone with basic math skills should be able to calculate his or her own experience mod. However, the employer does not know all of the information needed to calculate the experience mod without the experience rating worksheet used by the rating organization, That information includes the expected loss rate, D-ratio, credibility factor, and tabular ballast value. Since each rating organization may use a different format to present this information, it is suggested that an explanation of the worksheet used be requested.

5.2.4   Applying credits and debits

In addition to the classification rates which reflect the risk associated with various types of work, and the experience rating system which forecasts future losses based on past loss experience, premiums can be greatly affected by the often subjective process of applying various credits and debits. Once the manual premium has been calculated and modified through the experience rating process, credits and debits are applied to it. Similar to the experience mod, a credit implies a reduction in the premium, whereas a debit implies a premium increase.

Premium discounts are volume discounts — the larger the modified premium, the larger the discount — and are credits because they reduce the size of the premium. With the exception of minimum premiums, which are determined by each carrier, there is no such thing as a premium debit. The reason that companies with large modified premiums are rewarded by the application of a premium discount is because overhead expenses such as underwriting, loss control, managed care administration, and premium auditing do not increase proportionally with the premium but are nevertheless included in the earliest phase of the premium determination process as an element of the classification rate.

Whereas premium discounts are affected only by having a relatively large modified premium, other discounts are available to virtually all companies based upon their individual characteristics. Some of the things taken into consideration when applying these additional discounts are the existence of formal safety programs, employee selection criteria, employee training, drug-testing programs, modified duty programs, and even the carrier's perception of management's cooperation. Unlike premium discounts, the carrier may either credit or debit a premium based on these characteristics. Much of the information used to determine the applicability of credit or debit adjustments to the premium for these and other characteristics comes from the loss control representatives who visit companies prior to the carrier's initial quote of a premium or prior to a carrier renewing a policy. Even the observations made by the loss control representatives regarding the physical condition of the workplace and the actions of the employees during an onsite visit can influence the underwriter's decision to apply a credit or debit.

Whereas the above-mentioned credits and debits that are applied are primarily subjective judgments made by the carriers' underwriting staff, several state workers' compensation laws include legislated credits that are to be applied whenever certain actions are taken. For example, legislation in Pennsylvania provides a potential 5% premium discount if the employer implements a certified workplace safety program. In New Hampshire, employers who are enrolled in a managed care program are entitled to receive a 10% premium reduction. Similarly, legislation in Massachusetts provides for a 15% reduction in premiums for employers who are in the assigned risk pool and hire a qualified loss management firm to help them contain costs. Furthermore, Colorado employers who implement the specified steps in the state's Cost Containment Certification Program and receive safety certification become eligible for dividends of 2–10% and a 2.5% premium discount if they implement a designated medical provider program.

In addition to premium discounts, subjective credits and debits, and legislated credits, many insurance carriers have agreements with individual trade associations to provide a standard premium discount (for example, 10%)

for employers in that association. This discount is offered in exchange for the trade association's endorsement of the carrier as the provider of choice.

Regardless of the reason it is applied, a credit or debit applied to the modified premium can have a profound impact on the net premium ultimately paid by the insured.

Figure 5.3 shows a worksheet that includes all three elements of the premium calculation process. The worksheet illustrates a company that has made significant strides to address the cost of workers' compensation premiums and, as a result, the company has a net premium more than 35% less expensive than the manual premium. This company is paying considerably less in workers' compensation premiums than its average competitor, even though its modified premium was not significant enough to score a premium discount.

Premium Calculation Worksheet			
Company: XYZ Lumber and Building Supplies, Inc.			
Address: P.O. Box 4399, Awatha, Nevada 57104-4399			
Contact Person: Mr. Robert Swafford		**Telephone Number:** (630) 555-3763	
Classification	**Payroll**	**Rate (Per $100)**	**Premium**
Clerical	$100,000	$0.28	$280
Inside Sales Clerk	$250,000	$2.08	$5,200
Lumber Yard and Drivers	$250,000	$4.34	$10,850
		Manual Premium	$16,330
		Experience Mod	0.84
		Modified Premium	$13,717
		Premium Discount	0%
		Association Credit	10%
		Scheduled Credits	15%
		Total Credits	25%
		Net Premium	$10,288

Figure 5.3

5.3   Influencing workers' compensation premiums

Regardless of the industry, controlling the cost of necessary expenses, such as workers' compensation insurance, can constitute the competitive advantage needed for business survival and favor. However, many companies wait until their experience mod is a debit mod before tackling the overwhelming cost of workers' compensation insurance. The truth is that any company that is not systematically and zealously addressing the factors that influence its experience modification factor is paying too much for workers' compensation insurance and will continue to pay too much.

This section presents practical strategies that enable employers to reduce their workers' compensation premiums. Strategies include shopping the competition, actively seeking credits, ensuring that employees are properly classified, ensuring that incurred loss and premium information is accurately recorded, limiting indemnity losses, submitting informational claims, and reducing injury rates.

5.3.1 Reduce injury rates

The key to any long-term reduction in workers' compensation premiums is a single-minded focus upon employee safety and injury prevention. As presented throughout this chapter, premiums are largely based on the long-term success of injury prevention efforts. Although this text is intended to help employers address the high cost and challenges associated with managing workers' compensation, it is notable that the benefits of an effective safety program extend far beyond the parameters of workers' compensation. Even conservative estimates indicate that the hidden costs of employee injuries are four to six times greater than the amount paid in workers' compensation benefits. These costs include such things as lower productivity, decreased employee morale, increased employee turnover, and increased costs associated with human resource functions such as advertising, interviewing, and training.

To adequately address occupational safety and injury prevention, the company must make a commitment to safety beyond mere rhetoric, truly creating an atmosphere in which the safe performance of job duties is valued and is a condition of employment. This commitment to safety must begin with the senior administrators and upper management personnel and flow downward. Administrators should be involved and genuinely interested in the continuous improvement of the company's safety program. Hence, each member of the company's upper echelon should be informed of every injury shortly after it occurs and of the efforts undertaken to prevent future similar occurrences. Management personnel and line supervisors should be held accountable for the injuries within their departments and should be rewarded for their efforts that lead to an accident-free environment. Employers must treat employee safety an intrinsic element of each job function, not as a separate entity that must be continuously justified. Furthermore, senior administrators and upper management personnel must create a culture that pervades the entire organization, holding that accidents are unacceptable and can be prevented. It is only then that occupational safety and accident prevention will be perceived by the line employees as more than a collateral function.

Although the company that implements and maintains an effective safety program is the ultimate beneficiary, the company's workers' compensation carrier has a vested interest in preventing the injury of its clients' workers. For this reason, each workers' compensation insurance carrier employs loss control representatives who are responsible for providing the assistance and feedback that employers need. This assistance can take several forms, including the following:

- Motivating upper management personnel to embrace accident prevention as a company value.
- Providing recommendations for improving the company safety program.
- Participating in company safety meetings.
- Providing literature and training material.
- Assisting with the development of formal safety programs.

Employers should be cognizant that whether or not they use the services, they shoulder the insurance carriers' cost of employing their loss control representatives.

Other sources of assistance include the state and federal OSHA. With the intention of reducing work-related injuries, many of these organizations have separate divisions funded solely for the purpose of assisting employers achieve compliance with safety regulations. These OSHA divisions are generally separate from the compliance divisions and do not have the authority to issue citations for noncompliance. Similar to the insurance carriers' loss control representatives, these OSHA representatives are effactually safety consultants for whom the company is already paying through state and federal tax revenues.

5.3.2   Shop the competition

In addition to ardently addressing injury prevention, shopping the insurance market for competitive quotes can favorably influence workers' compensation premiums. In a free-market economy, competition breeds lower costs. For this reason, employers are encouraged to determine which carriers provide workers' compensation coverage in the state and to receive as many quotations for coverage as possible. This process is somewhat complicated by the fact that some insurance agents work for a single carrier and thus provide quotes for only that carrier. Independent agents can obtain quotations from a number of different companies. However, even among independent agents, some have agreements with insurance carriers who limit the number of agents with whom they do business. The reason for these agreements is simple. Since developing a quote consumes an underwriter's time, it is not cost-effective for insurance carriers to develop premium quotes for independent agents who very seldom place business with them. In short, employers must do their homework to determine which carriers offer the most competitive rates and then ensure that the agent with whom they are working is able to provide quotes from those carriers.

5.3.3   Actively seek credits

It is not sufficient to seek the carrier with the lowest classification rate because a carrier with low classification rates may not be willing to apply significant credits. However, to receive credits, an employer must first know the things for which the carrier is willing to offer credits. These often include premium discounts, affiliation credits, and credits based on the individual characteristics of the employer.

The employer has little control over whether or not a premium discount will be applied because the premium discount is based solely upon the amount of the employer's modified premium. However, some carriers may be willing to apply a higher percentage of premium discount to the modified premium than others.

An affiliation credit is often applied to the premium of employers who are members of a trade association. When trade associations enter into agreements with workers' compensation carriers, they are rewarded for their endorsement of the carrier through monetary compensation, often in the form of a percentage of the net premium paid to the carrier by each association member. For many trade associations, it is a major source of revenue. Consequently, many trade associations have agreements with a particular carrier. The first step for employers to receive this type of credit is to contact each of the trade associations of which they are members and to inquire if such an endorsement agreement exists and what economic advantage exists. If the employer is not a member of trade association, or if its trade associations do not have such an endorsement agreement with a particular workers' compensation carrier, the next step is to contact other trade associations closely affiliated with the employer's industry. Frequently the cost of membership in an association is less than the credit extended from the insurance carrier with whom an endorsement agreement has been made. For this reason, it may be cost-effective to join a particular trade association simply for the workers' compensation premium reduction.

In addition to premium discounts and affiliation credits, a wide variety of credits are offered based on individual characteristics of the employer. These characteristics fall into the following categories: formal safety programs, cost control efforts, management involvement and cooperation, employee selection and training, and the care and condition of the premises.

Common to almost all carriers is their appreciation and recognition that formal safety programs have a positive impact upon the prevention of injuries. Hence, a company that has gone through the sometimes arduous task of creating, implementing, and maintaining formal safety programs is rewarded for its efforts, not only by the resultant reduction in injuries, but also by a credit to the modified premium. Safety programs include those mandated by OSHA, such as a Hazard Communication Program, Lockout/Tagout procedures, an Emergency Action Plan, and an Exposure Control Plan for Bloodborne Pathogens. Formal safety programs also include things such as routine safety meetings, a safety committee, accident investigations, posted safety rules, formalized disciplinary procedures, and safety incentive programs.

Workers' compensation carriers recognize that not only formal safety efforts affect the potential for future injuries, but that individual employees do as well. For this reason, credits may be available for companies with measures intended to attract desirable workers, dismiss undesirable workers, and maintain a stable workforce. These measures may include requiring successful applicants to possess a specified amount of past work experience, reviewing the work experience of applicants prior to hire, conducting pre-placement drug-testing and physical examinations, conducting random drug/alcohol testing of employees, and terminating employees who fail to abide by established safety procedures.

Insurance carriers are also pleased to see employers who have implemented measures to control the cost of accidents after they occur. These measures include the consistent use of modified duty assignments to eliminate or lessen the indemnity losses of individual claims and use of a managed care program and/or onsite medical care for small injuries. When these cost control strategies are implemented, the insurance carrier recognizes that even when losses occur, the severity of the loss will likely be minimized by the efforts of the employer. As such, many carriers are willing to offer percentage credits for such initiatives.

Not only are insurance carriers concerned with the initiatives that have been undertaken thus far to limit the frequency and severity of losses, but they are also interested in the willingness of the employer to implement such processes. The perceived cooperation of the employer can be quite influential when the time comes for the underwriter to determine whether or not to apply a credit. For this reason, it is essential for employers to convey to the agent and to any individual representing a specific insurance carrier a willingness to implement recommendations. Frequently these recommendations, as well as the insurance carrier's perception of the employer's cooperation, are a result of meeting with a loss control representative from the insurance carrier. These loss control representatives are charged with the dual responsibility of painting an accurate picture of the employer's operation to the underwriter and assisting the employer to achieve the goal of lower accident frequency and severity.

Closely related to the employer's perceived cooperation with the carrier is the carrier's perception of the employer's involvement and attitude relative to employee injuries. An employer who holds the view that injuries happen and there is little that can be done does not portray a desirable image of the company. Conversely, an employer who has personal knowledge of each injury and what has been done to prevent future similar injuries presents a quite different image. Hence, when meeting with loss control representatives, the employer should be cognizant that the brief time spent with that representative could make a difference.

Yet another characteristic of individual employers that can create a credit is the care and condition of the facility in which employees work. Generally this subjective judgment is made by the loss control representative. Whereas it is not recommended that employers simply clean up the facility when there is an announced visit from the insurance carrier's loss control representative, it is important to be aware that, fair or not, the appearance of the facility helps form the carrier's perception of the operation. Good housekeeping, orderly maintenance of stock, adequate lighting, and general facility upkeep create the image of a company that is well-organized and not too busy with production to be concerned with collateral duties that may affect accident frequency.

In short, an insurance carrier is taking a risk when it insures any individual employer. The goal of the carrier is to use as much relevant information as possible about the risk to accurately gauge the degree of risk. A company that actively seeks credits is well-advised to take the time in preparing a profile of their company which describes each of the aforementioned characteristics. The goal of the employer is to address any concerns that the carrier might have concerning the potential for loss. Hence, this profile should address formal safety programs, cost control efforts, management involvement, employee selection and training, and the cooperation that can be expected from the employer. This profile might also include a commentary relative to past losses, indicating the measures that have been taken to prevent future similar occurrences. The latter is particularly relevant if there has been one or more major losses or a trend of injuries. A sample profile is provided in Appendix C. This profile was written to a prospective insurance carrier to familiarize the carrier's underwriting staff with the efforts that had been undertaken to address employee safety, accident prevention, and the control of workers' compensation costs.

5.3.4   Ensure that employees are properly classified

As explained earlier in this chapter, the manual premium is calculated by multiplying the payroll in each classification by the rate charged for that classification. The rate is based on the potential for loss. If an employer reviews the audits and determines that one or more employees have been improperly classified, the employer may be paying too much for coverage and may be entitled to a refund. To address the potential for misclassifications, employers must be familiar with the classification descriptions that most closely resemble the work that employees perform. To achieve that familiarity, the employer should obtain a copy of the manual that describes the classifications and use it as a reference following each premium audit.

One thing that should cause employers to scrutinize the premium audit more closely is a premium audit classification that is not reflected on the workers' compensation policy at the beginning of the policy year. Whereas this appearance could be the result of a change in the operation since the issuance of the policy, it could be an unscrupulous attempt to place employees into a classification with a higher rate.

As discussed earlier, the classification of employees is not an exact science. Instead, employees are placed into the most appropriate classification based on the knowledge of the operation held by the insurance agent, the insurance carrier, and the premium auditor. If the individuals who make the classification determination do not have firsthand knowledge of the operation, entire groups of employees can be improperly classified. Such misclassifications can go unnoticed for years because since most employers rely solely upon their agent and insurance carrier to ensure proper classification.

Another situation in which a portion of the payroll could be improperly classified is the change of duties of one or more employees during the policy year. If this occurs and the premium audit does not separate that payroll into two separate classifications, the employer could be paying a higher-than-necessary rate. An example of this is an employee who is a production worker at the beginning of the policy year but is promoted to an administrative position during the year. If the rate for a production worker is $7.85 per $100 in payroll and the rate for an administrative (clerical) employee is $0.28 per $100 in payroll, it is clear that a significant premium overcharge would result if the employee's payroll were calculated in the production classification for the entire policy year.

Yet another situation that can cause misclassification of employees is the fact that classifications sometimes change. For example, a classification may initially include both drivers and production workers for a particular industry within one classification. However, a change may separate the drivers and the production workers into two separate classifications, with different rates. If the premium audit does not recognize the revised classification description and separate drivers and production workers accordingly, several employees will be improperly classified.

Last, consider all employees whose job duties are divided among completely different functions, such as a truck driver who also works as a dispatcher. Generally insurance carriers will place an employee in the classification with the highest rate. However, if the employer is able to accurately separate the payroll of employees who perform two or more separate job functions, the workers' compensation carrier may be amenable to placement of the employee into two or more classifications by splitting the individual's payroll according to the time spent performing each job. This can make a substantial difference in the premium since the rates can vary significantly. For example, the rate for a truck driver may be $15.00 per $100 of payroll, whereas the rate for a dispatcher may be $0.28 per $100 of payroll.

5.3.5   Ensure that premium information is accurately recorded

Since the payroll information from which premiums are calculated is based on the employer's estimation of payroll for the upcoming year, a premium audit is conducted following the conclusion of the policy period to determine the actual payroll during the previous policy year. It is essential that this information be accurate since inflated payroll translates into inflated premiums. During the premium audit, many companies inadvertently overstate their payrolls by including remuneration, which is excludable. Tips, overtime, severance pay, expense reimbursements, and pay for active-duty military service are often excludable from the payroll audit. Additionally, there are limits on the reportable wages earned by corporate officers. If this information is included in the remuneration, the company will pay too much for its workers' compensation insurance.

Since the individuals who conduct premium audits are human, they are capable of oversights and mistakes. For this reason it is essential to review the payroll information to ensure that it is accurate. Furthermore, some insurance carriers permit the companies that they insure to audit themselves. For these companies it is even more important that they understand what remuneration is to be included and what can be excluded. Since these rules vary with each state, the employer must become familiar with the remuneration guidelines in the states in which it does business.

Although overstated payroll during a premium audit causes an employer to pay too much for workers' compensation insurance, the reverse is true with payroll information used to calculate an experience mod; understated payroll can cause an experience mod to be improperly inflated. To understand this, consider a company with $500,000 in payroll in a given classification. If the data were entered incorrectly as $50,000 when calculating the experience modification factor, the result would be an inflated experience mod since the expected losses would be based upon a premium of $50,000 instead of $500,000.

5.3.6 Ensure that loss information is accurately recorded

Just as undervalued premiums used in the experience rating process can increase the experience mod, improperly recorded loss information can also result in an inaccurately high experience mod.

To ensure that loss information is accurate, the employer should get a copy of its loss runs from each workers' compensation carrier through which it maintained coverage during the three-year period used to calculate the experience mod. Loss runs are documents used by each carrier to display each of the loss events incurred during the period of coverage and the corresponding amount incurred for each loss. Loss runs include both the amount paid to date and the amount which has been reserved for that loss. When the carrier believes that it will incur no more cost for a particular claim, the claim is termed "closed" and the paid-to-date amount is equal to the reserve amount. If the carrier believes that more costs may be incurred by a claim, it is termed "open" and a reserved amount greater than the amount paid to date is indicated. This reserved amount is generally considerably higher than the amount that will ultimately be incurred; it is reserved as an accounting method to keep money set aside for paying the costs of each claim.

Ideally, loss runs should be obtained prior to the insurance carrier's submitting information to the rating organization. The employer is then able to address any concerns before the information is used to calculate the experience modifier. These concerns include ensuring that each of the claims listed on the loss runs belongs to the proper employer and that open claims are justifiably being left open.

Although not a frequent problem, loss information is occasionally assigned to the wrong company, often the result of a simple data entry mistake by the insurance carrier. When this situation occurs, it can generally be corrected with a quick telephone call to the insurance carrier informing them of the mistake. However, if the mistake is not identified until after the information has been sent to the rating organization and an experience mod has been calculated for the upcoming policy year, the process for correcting the problem becomes more complicated.

A more frequent source of poor loss information used to calculate an experience mod is the carrier's failure to close a claim before reporting information to the rating organization. To illustrate this situation, consider a single loss event for which the insurance carrier has established a reserve of $50,000. If the loss incurred only $12,500 and there is no reason to believe that any more will be incurred as a result of that loss, the claim should be closed. If left open, a $50,000 loss will be used in the calculation of the experience mod, compared with $12,500 if the claim were closed prior to the information being sent to the rating organization.

Once the experience rating worksheet is received from the rating organization, the employer should compare the loss information from the loss runs, including any changes that have been made, with the loss information on the experience rating worksheet. To make this comparison, the employer should use the split-rating system described in a previous section of this chapter, separate each of the losses into primary and excess losses, and then compare that information to the information used on the experience rating worksheet.

5.3.7   Limit indemnity losses

Indemnity losses are the amounts paid to claimants as compensation for lost wages. Frequently they can exceed medical treatment and rehabilitation costs. For example, consider a truck driver who fractures his left ankle and is unable to drive for six weeks. The medical cost for this claim may be $1500 or less. However, the indemnity benefits paid to the employee for lost wages may be $2500. Since the calculation of the experience mod does not take into consideration whether the loss amount is a result of medical costs or indemnity benefits, the claim is recorded as a $4000 loss. However, if the company were able to eliminate the indemnity benefits associated with this claim, the total cost of the claim would be reported as $1500. This difference has a significant impact on experience mod calculation.

The most effective means of limiting or eliminating indemnity benefits is to develop and implement an effective modified duty program, a topic treated in depth in Chapter 7.

5.3.8   Submit informational claims

Most employers are instructed by their workers' compensation carrier to submit a report of injury for every work-related injury. Whereas this text does not propose to contradict the policies of individual insurance carriers, it is important to remind employers that the frequency of losses has a significant impact on the experience mod. This quandary lends itself to a solution that many carriers accept as common practice, the submission of "information-only claims."

When a company submits an information-only claim, it is effectually informing the insurance carrier that an injury has occurred but will not result in any cost to the insurance carrier. It is typically done for relatively small injuries, which the employer pays through the company budget, instead of submitting all claims to be paid by the carrier. In such circumstances, the insurance carrier will accept a report of injury for the claim, annotated with the message that it is an "information-only claim." The carrier records the information but does not pay any of the costs associated with the claim. Since neither the employer nor the carrier can look into a crystal ball and determine if a claim will develop into a more serious and costly loss than originally thought, the carrier will often allow the employer to reconsider the decision to pay for the claim internally. At that point, the insurance carrier will assume the claim and begin paying the bills.

Whereas these information-only claims will likely be displayed on the loss runs received from the insurance carrier, they do not affect the experience mod, assuming that the employer does not reconsider his decision and have the carrier assume payment. As illustrated in the above explanation of how the experience mod is calculated, the number of claims is not used in the formula.

5.4   Summary

It is impossible for a company to have a significant, favorable, and lasting impact on the cost of workers' compensation coverage if the individuals within the company who have that responsibility do not have a sound understanding of the factors that influence premiums. To understand these factors, it is beneficial to have a fundamental knowledge of the premium calculation process.

In brief, workers' compensation premiums are calculated in a three-stage process, which involves determining the manual premium, adjusting the manual premium by a process known as experience rating, and then applying miscellaneous credits and debits. During each of these stages within the premium determination process, the employer can favorably affect the net premium. To effect a lower manual premium, the employer can shop the competition for the lowest rates, ensure that employees are properly classified, and determine if any employee in a higher-rated classification can be separated into a lower-rated classification for the portion of his or her job that involves less hazardous work. To facilitate premium reduction during the experience rating stage, the employer can reduce injury rates, limit indemnity expenses, ensure that the information used to calculate the experience mod is accurate, and submit informational claims. Additionally, employers can have a profound impact upon lowering their workers' compensation premiums simply by knowing the characteristics for which individual insurance carriers are willing to offer credits. These often include the implementation of formal safety programs and cost control efforts, the existence of management cooperation, effective employee selection and training, and maintaining the proper care and condition of the premises.

chapter six

The workers' compensation claims coordinator

Contents

6.1   Introduction

Similar to other functions within a business, controlling the costs relating to workers' compensation requires effective management. Other functional divisions of businesses such as purchasing, marketing, production, and the like have assigned managers, supervisors, or coordinators. In fact, few business owners would feel comfortable without a single individual to hold accountable for these operational functions. Therefore, if a company is truly sincere about desiring to control the costs relating to workers' compensation, an individual employee should be designated to manage that function, as well.

Despite the fact that workers' compensation costs must be managed similarly to any other functional division of a company, many companies divide the responsibilities relating to workers' compensation claims management among several individuals. A company may assign the responsibility for completing the First Report of Injury or Illness Form to the injured employee's supervisor. Reporting the claim to the insurance carrier may be the responsibility of a clerical employee. Furthermore, the review of loss runs may be something that is done by the finance department. This practice divides the management of workers' compensation costs into the performance of specific tasks and thereby compartmentalizes the overall function of workers' compensation claims management. One problem with this approach is that the specific functions that relate to workers' compensation claims management are very much interrelated and are thus most effectively carried out, or at least overseen, by one person who has firsthand knowledge of what has been done and what needs to be done to manage specific claims effectively. Another problem is that compartmentalizing the functions of workers' compensation claims management assumes that the effective management of workers' compensation claims requires minimal, if any, knowledge and understanding of the overall process, that it requires no working relationships to be developed, and that it requires no real management skills. Yet another problem with this approach is that personnel changes among any of the individuals who have been assigned specific tasks can disrupt the entire process.

What is even less effective than compartmentalizing the management of workers' compensation claims is to say that controlling workers' compensation costs is the responsibility of all employees. This is nothing more than well-intentioned rhetoric. A company that fails to assign responsibility to an individual for any management function and proclaims that function to be the responsibility of all employees negates the ability to hold individuals accountable for shortcomings. Additionally, it is likely that this approach will eventually lead to finger-pointing — "I thought he was going to do that."

It is necessary, then, for every company to assign responsibility for the management of workers' compensation claims to an individual employee. The process of effectively managing workers' compensation claims entails a multitude of interrelated tasks and requires the responsible employee to gain knowledge and develop effective working relationships with the other players in the workers' compensation process. For these reasons alone it is essential.

This text uses the title "workers' compensation claims coordinator" to describe the person to whom these responsibilities are to be assigned. The intent of this chapter is to define the term and describe the functions with which the workers' compensation claims coordinator should be charged. Additionally, this chapter provides some guidelines for the selection of a workers' compensation claims coordinator.

6.2   What is a workers' compensation claims coordinator

A workers' compensation claims coordinator is an employee assigned the overall responsibility of controlling the direct and indirect costs relating to workers' compensation claims.The workers' compensation claims coordinator is the adhesive that holds together the company's workers' compensation cost control efforts and serves as the source of information regarding workers' compensation for employees and management. Furthermore, the workers' compensation claims coordinator is the point of contact for all outside sources concerning workers' compensation claims. Without the concerted effort and effective leadership of the workers' compensation claims coordinator, the success of a company's efforts to control workers' compensation costs through effective management is doomed from the start.

Although the duties of a workers' compensation claims coordinator are extremely important, it is not a full-time position for most companies. The position is rather a collateral function of an employee's primary job. In other words, the workers' compensation claims coordinator title is but one of the hats worn by this employee.

6.3   Selecting a workers' compensation claims coordinator

Because the workers' compensation claims coordinator is intimately involved in every aspect of workers' compensation, a component of business which can have a profound impact on the company's bottom line, due diligence should be practiced in the selection of the employee assigned to this position. Although ill-advised, the tendency may be to give little consideration to whom the responsibilities of the workers' compensation claims coordinator are given.

As evidenced by the information contained in this chapter, the workers' compensation claims coordinator has myriad responsibilities. In this age of downsizing and corporate restructuring, many managers already have a full plate. Selecting an employee who already has a full workload, without relieving some existing responsibilities, will undoubtedly lead to some degree of neglect of one or more of those responsibilities. In short, there is a limit to what can be done well by one person. For this reason, the responsibilities of the workers' compensation claims coordinator should be reviewed, and a weekly or monthly time allotment should be estimated for these duties and considered when selecting a workers' compensation claims coordinator.

Whether or not a company has given any specific employee the title of safety director, most companies have one person who has overall responsibility for the administration and management of employee safety and health efforts. Because of the relatedness of employee safety and workers' compensation, the employee charged with the administration and management of employee safety and health is likely to be the employee who is best suited for the responsibilities of workers' compensation claims coordinator. Both positions require intimate knowledge of all work-related injuries, the measures taken in response to those injuries, and the steps taken to prevent future similar occurrences. Hence, assigning someone other than the safety director to the position of workers' compensation claims coordinator may result in a duplication of job duties.

Although the company's safety director may be the most logical choice, other employees should not be overlooked. In every company there exists at least one employee who is already involved in the workers' compensation process, such as the person who currently submits claims to the workers' compensation insurance carrier or, for large companies, an onsite occupational nurse or physician. Because of their knowledge, experience, and involvement with the workers' compensation process, these individuals should also be considered as potential candidates for the workers' compensation claims coordinator.

Regardless of who is selected as the workers' compensation claims coordinator, he or she must possess or be granted authority commensurate with the magnitude of responsibility. This position involves interacting and providing instruction to employees at all levels of the organization, including supervisors, managers, and company executives. As such, the person selected for this position must be able to effectively perform the functions of the position without being hindered by uncooperative managers who exercise their authority over the selected candidate. For this reason it is not recommended that the person charged with these responsibilities be a line employee who has no management authority.

In addition to the existing duties and authority of employees, the personal traits of individuals considered for the postion of workers' compensation claims coordinator should be contemplated, remembering that the selection of the appropriate candidate can impact the company's bottom line in either direction. Traits to consider include existing knowledge, professionalism, empathy, written and oral communication skills, and organizational skills.

Selecting an employee with existing knowledge of workers' compensation can make the move to a formal workers' compensation management program a much smoother transition. Although background relative to medical terms and treatments is beneficial, it is not essential, as most medical professionals and insurance representatives can be quite helpful in explaining specific concerns as they arise. Similarly, knowledge of workers' compensation insurance in the states in which the company operates is beneficial but can be learned through involvement in the process and a willingness to ask questions.

In addition to existing knowledge, the person selected as the workers' compensation claims coordinator should be professional. In this context, professionalism is an attribute that includes a demeanor that elicits respect. The selected candidate will be communicating with physicians, nurses, insurance representatives, attorneys, and other professionals outside the organization, and he or she must do so effectively to provide the best possible outcome for the company. A lack of professionalism is often easy to detect and may hinder the ability to develop effective working relationships with these other professionals. Furthermore, an employee whose attitude and actions elicit the respect of line employees and managers will be better able to acquire cooperation of persons within the organization.

Closely related to the attribute of professionalism is the motivation of the selected candidate. The employee selected to serve as workers' compensation claims coordinator should be goal-driven. The control of workers' compensation costs is an ongoing process that is never fully completed. An employee who is goal-driven will be motivated to set and achieve goals directed at the containment of costs relating to workers' compensation. In contrast, an employee who is apathetic and views work-related injuries and abuses of the workers' compensation system as inevitable will undermine the company's efforts to control workers' compensation costs.

Another personal trait to consider when selecting a worker's compensation claims coordinator is empathy. It is stressed throughout this text that employees who are injured in the course of employment can often become terrified of the uncertainties that lie ahead, not knowing whether their medical bills will be paid, if they will have a job when released to return to work, or if they will be viewed differently by their supervisor or peers as a result of their misfortune. With all of this internal turmoil, the workers' compensation claims coordinator must be able to display a genuine concern for the injured employee, empathize with the injured employee, and convince the employee that the company is an advocate, as opposed to an enemy, for injured workers.

Well-developed interpersonal communication skills are also an essential attribute for an effective workers' compensation claims coordinator. This position will require a significant amount of oral communication within and outside the company. Within the company, the workers' compensation claims coordinator will be responsible for conducting training of line employees and managers, conducting accident investigations, and soliciting the cooperation of other managers and supervisors in assigning modified duty tasks. Additionally, the workers' compensation claims coordinator will be involved in direct communication with injured employees and their family members. Individuals outside the organizations with whom communication is common include physicians, insurance claims representatives, and attorneys. Effective communication with these outside entities is essential ensure proper treatment, utilize modified duty to its fullest extent, ensure the prompt payment of legitimate claims, deny noncompensable claims, and avoid litigation.

In addition to oral communication skills, the candidate selected for the position of workers' compensation claims coordinator should possess good written communication skills. In short, the employee selected for this position cannot be someone who despises paperwork. This employee will be required to submit First Report of Injury Forms and supporting documentation to the insurance carrier, will be instrumental in the development of a written Modified Duty Program, will maintain documentation relating to the physical requirements of jobs, and may be required to initiate and respond to written correspondence with the treating physician, the insurance carrier, and attorneys. Additionally, this employee will be required to initiate documentation in a manner that affords the ability to evaluate the overall success of the workers' compensation management efforts.

Because of the amount of documentation relating to workers' compensation and the myriad responsibilities of the workers' compensation claims coordinator, it is essential that the employee selected for this position possess good organizational skills. Depending on the size of the operation and the number of workers' compensation claims submitted, the workers' compensation claims coordinator may be juggling several different claims simultaneously. Coupling this with the routine responsibilities of the selected employee (those unrelated to workers' compensation), it is easy to see that well-developed organizational skills are imperative.

With all of the characteristics and personal attributes desired, it may seem as if the ideal candidate for the position of workers' compensation claims coordinator is every employer's dream. In fact, the ideal candidate may not exist. However, it cannot be overemphasized that the selection of the most competent employee is pivotal to the success of workers' compensation cost control efforts.

6.4   Roles and responsibilities of the workers' compensation claims coordinator

As evidenced by the following description of duties, management efforts intended to control the costs of workers' compensation include a wide range of activities. As presented, these duties are the responsibility of the workers' compensation claims coordinator. However, depending on that person's workload, several of these duties may be delegated to other responsible individuals within the organization. Nevertheless, the person given the workers' compensation claims coordinator title is the sole person charged with coordination of these activities and the person who should be held accountable for achieving related goals.

6.4.1   Coordinating the development and implementation of relevant policies and procedures

Similar to other goal-driven endeavors within a company, efforts that are intended to control the costs of workers' compensation are routinely directed by company policies and procedures. However, to be effective, an individual with the requisite knowledge must coordinate the development of these policies and procedures. If due diligence has been exercised when assigning an employee to the position of workers' compensation claims coordinator, then it is that person who is likely the most appropriate employee to coordinate the development and implementation of relevant policies and procedures.

Among the policies and procedures that should be developed to achieve the desired cost control objectives are those which address accident reporting, use of modified duty, accident investigations, detection and deterrence of workers' compensation fraud, employee training, and record-keeping. Since the policies and procedures that address these issues have interrelated elements, it is necessary that the workers' compensation claims coordinator create the guidelines in a manner that avoids unnecessary duplication of efforts.

Since the management of occupational safety and health is so closely related to the efforts to control workers' compensation costs, the line that divides the two is often blurred. Therefore, even if the designated workers' compensation claims coordinator is not the same person who is responsible for managing the company's occupational safety and health efforts, he or she must not be excluded form the development and evaluation of policies intended to address safety issues. Consequently, it is not only the responsibility of the workers' compensation claims coordinator to coordinate the development of the aforementioned policies and procedures, but also to be involved in the development and evaluation of policies and procedures that address issues such as safety inspections, preventive maintenance programs, job hazard analyses, use of personal protective equipment, safety meetings, and formal programs mandated by OSHA.

6.4.2   Educating and training employees, supervisors, and managers

Although the efforts aimed at controlling workers' compensation costs are coordinated by the workers' compensation claims coordinator, it is important to note that all individuals have defined responsibilities. These responsibilities are likely included in the aforementioned policies and procedures. However, it is the responsibility of the workers' compensation claims coordinator to inform line employees, supervisors, and management personnel of their respective roles. Without an effort to educate and train all employees, it is likely that the intended utility of the policies and procedures will not be realized.

In addition to conducting training that identifies the specific roles of individuals and groups of employees, the workers' compensation claims coordinator should conduct training which informs all employees of the intent of the policies and the impact that increased workers' compensation costs could have, or have had. Furthermore, training for management and supervisory personnel should emphasize their ability to influence and motivate subordinate employees purely through their visibly evident awareness of the significance of controlling workers' compensation costs.

6.4.3   The central contact person for all work-related injuries

The workers' compensation claims coordinator should be the central contact person for all work-related injuries. An ideal practice is for the respective supervisor to accompany the injured employee to the workers' compensation claims coordinator. However, with loss events in which immediate medical attention is warranted, the supervisor of the injured employee should obtain as much information as possible surrounding each work-related injury as soon as it occurs and then provide that information to the workers' compensation claims coordinator without delay. By serving as a central contact person, the workers' compensation claims coordinator is able to ensure that all post-accident procedures are being properly executed. These procedures may include having a supervisor accompany the injured employee to the medical provider, ensuring that an accident investigation is initiated, ensuring that the events surrounding the accident are evaluated for evidence of fraud, and ensuring that the injured employee and the supervisor are familiar with their respective responsibilities.

6.4.4   Submitting claims to the workers' compensation insurance carrier

Because the workers' compensation claims coordinator serves as the central contact person for all work-related injuries, it is logical that that person be responsible for submitting necessary information to the workers' compensation insurance carrier. As such, the workers' compensation claims coordinator must be familiar with the preferred means of reporting injuries. Although all workers' compensation insurance carriers require a specific injury report form to be completed, some prefer the information to be faxed while others have electronic means of reporting the information. Still others may accept claim information through regular mail or via telephone reports. Regardless of the media used when transmitting information to the insurance carrier, the workers' compensation claims coordinator should retain copies of the documentation sent.

When submitting initial claim information to the insurance carrier, the workers' compensation claims coordinator must indicate whether the information is being submitted as an actual claim to be paid by the insurance carrier, or if the claim is being submitted as information-only, for which the employer will bear the cost of the medical expenses. This practice is done to keep claim costs to a minimum and is described further in Chapter 5. However, for the purpose of identifying the duties of a workers' compensation claims coordinator, it is necessary to recognize that the workers' compensation claims coordinator, must possess the knowledge and authority to differentiate between claims for which the company intends to retain financial responsibility vs. those for which the insurance carrier will be expected to absorb the costs.

6.4.5   Screening claims for evidence of fraud and claims of questionable compensability

In addition to serving as the central contact person for work-related injuries and reporting claims to the insurance carrier, the workers' compensation claims coordinator should be held responsible for assessing each reported injury for evidence of possible fraud or questionable compensability. This can be done simply by using a checklist of red flag indicators and forwarding that information to the insurance carrier.

6.4.6   Coordinating and reviewing accident investigations

The topic of conducting accident investigations is discussed in Chapter 8 of this text and identifies criteria for selecting the most appropriate person within the company to conduct accident investigations. Whereas the most appropriate person for this responsibility may be the workers' compensation claims coordinator, it is not necessarily a required duty of that person.

Regardless of who is charged with the responsibility of conducting accident investigations, the workers' compensation claims coordinator should be charged with informing the employee who has been delegated the duty of conducting accident investigations when an accident occurs. Furthermore, it is the responsibility of the workers' compensation claims coordinator to review the accident investigation to ensure that it has yielded the necessary information.

6.4.7   Ensuring that corrective measures have been implemented

The workers' compensation claims coordinator ensures that corrective measures have been implemented following an accident investigation. These corrective measures should be addressed in every completed accident investigation. They include both interim and long-term corrective measures and address both surface and root causes.

6.4.8   Acquiring return-to-work documentation

Whenever an employee is injured and receives professional medical treatment, there are three possible immediate outcomes (assuming that the injury is not fatal or does not permanently disable the employee). These immediate outcomes are based on the seriousness of the injury and the treating physician's diagnosis for recovery.

One possible outcome is that the treating physician releases the employee to return to work immediately without any temporary physical restrictions. In such circumstances, the workers' compensation claims coordinator should obtain a written statement (commonly called a "doctor's release") from the treating physician and maintain that documentation.

Another possible outcome is that the treating physician releases the employee to return to work with temporary physical restrictions. These restrictions should be written and specific. They should indicate specific physical restrictions (such as "no lifting more than 50 pounds") and should specify the period for which the restrictions are valid (such as "10 days"). If the physician-imposed physical restrictions are not specific, the workers' compensation claims coordinator should be held responsible for contacting the physician immediately and requesting written clarification.

The third possible outcome is that the treating physician states that the employee should remain off work for a recovery period. Although this diagnosis does not appear to enable the use of modified duty, the workers' compensation claims coordinator should contact the treating physician immediately and advise that the company has a modified duty program and is willing to accommodate temporary physical restrictions during the employee's recovery period. This is one place that professionalism and good interpersonal communication skills may prove effective. Often, the willingness of an employer to be involved in the injured employee's recovery is sufficient reason for the treating physician to rethink his or her initial decision. However, it is still essential for the physician to provide specific written physical restrictions, as described above. It is notable that the physician will not always permit the employee to return to work during the recovery period and that a modified duty assignment should not be contemplated if the treating physician will not provide written temporary physical restrictions.

6.4.9 Coordinating modified duty strategies

Once the employee is released to return to work with temporary, physician-imposed physical restrictions and the written physical restrictions have been acquired, it is the responsibility of the workers' compensation claims coordinator to meet with the injured employee and the injured employee's supervisor as soon as possible to design a modified duty assignment that will conform to the employee's prescribed limitations. This will entail reviewing the physician-imposed, temporary physical restrictions and will also often involve reviewing previously compiled written job descriptions and potential modified duty tasks.

Again, good interpersonal communication skills on the part of the workers' compensation claims coordinator are essential, as competing ideas must be synthesized to develop an acceptable modified duty assignment. It may sometimes be necessary to involve other individuals within the company, such as supervisors from other departments, to help provide modified duty tasks that conform to the physician's temporary restrictions.

Through meeting together with the injured employee and the injured employee's supervisor (and perhaps other supervisors), a modified duty assignment should be developed and documented. The workers' compensation claims coordinator should maintain a copy of that documentation.

6.4.10   Maintaining effective communication with injured employees

One of the primary responsibilities of the workers' compensation claims coordinator is to maintain effective communication with injured employees. As previously mentioned, employees who sustain injuries at work are often apprehensive after they are injured. Because they are unfamiliar with the workers' compensation system, they may not know whether their medical bills will be paid, if they will have a job when released to return to work, or if they will be viewed differently by their supervisor or peers because of their involvement in an accident. Hence, as soon as possible following an injury, the workers' compensation claims coordinator should quell their fears by explaining what will transpire.

The goal of the workers' compensation claims coordinator during interaction with the injured employee is not only to make the employee feel more at ease but also to protect the employer by fostering a favorable employee/employer relationship. Depending on the workers' compensation claims coordinator's response to an employee's injury claim, a workers' compensation claim may be short-lived and have a relatively low severity (in terms of total cost) or may be much more costly than necessary, with the employee seeking to prolong the injury, obtaining a lawyer, or simply being uncooperative. Therefore, from the moment that an employee reports an injury until he has been released from the physician's care, every interaction with that employee can affect the employee's view of the company.

Although favorable interaction is critical when an injury is first reported, the workers' compensation claims coordinator must maintain effective communication with the injured employee throughout the period of treatment and recovery. If the employee is off work as a result of the injury, the workers' compensation claims coordinator should make frequent calls to the employee to ascertain progress and to communicate the company's concern for the employee's well-being. The workers' compensation claims coordinator may even coordinate assistance needed by the injured employee, such as grocery shopping or transportation to physician visits. If the employee has returned to work following a work-related injury, the Workers' Compensation Claims Coordinator's communication with that employee may simply involve routinely inquiring about medical progress, future physician visits, and any problems performing work as a result of the injury.

6.4.11   *Maintaining effective communication with physicians*

The workers' compensation claims coordinator's responsibility of maintaining effective communication is not limited to injured employees. He or she is also responsible for maintaining effective communication with physicians.

If the company has a limited number of physicians who treat injured employees, the workers' compensation claims coordinator should ensure that those physicians are familiar with the physical requirements of the jobs that employees perform. This communication may be facilitated through providing written job descriptions to the physicians or inviting the physicians to the company facility to see the work that is performed. Additionally, the workers' compensation claims coordinator should effectively communicate the company's policy relative to modified duty and should describe potential modified duty assignments. By doing these things, the physician will likely be more agreeable to permitting modified duty assignments for work-related injuries.

Even if the company does not have a limited number of physicians who treat injured employees, it is the responsibility of the workers' compensation claims coordinator to inquire about the employee's medical progress from the treating physician and to ascertain what role the employer might have in the injured employee's recovery. Communication with the treating physician will also enable the workers' compensation claims coordinator to determine if the injured employee is attending all of his or her scheduled medical or rehabilitative visits.

6.4.12   *Maintaining effective communication with insurance carrier's claims representatives*

Because the management of workers' compensation claims is a combined effort between the employer and its workers' compensation insurance carrier, it is imperative that the workers' compensation claims coordinator maintain constant communication with the claims representatives of its insurer. This includes communicating all details surrounding each loss event, the scheduled dates of medical treatments, and the results or findings of medical treatments. With respect to modified duty, the workers' compensation claims coordinator should immediately advise the workers' compensation claims representatives of the work status of each claimant in a workers' compensation claim. For the purposes of determining eligibility for indemnity benefits and to assist in the management of claims, the insurance claims representative must know whether the employee has returned to full-duty work, has been instructed by the treating physician (in writing) to remain off work, or has been instructed by the treating physician (in writing) to return to work with temporary physical restrictions.

In short, the relationship between the workers' compensation claims coordinator and the insurance claims representative should be viewed as a team managing workers' compensation claims.

6.4.13 Maintaining effective communication with company managers and owners

Not least important with respect to maintaining effective communication is the responsibility of the workers' compensation claims coordinator to communicate with company managers and owners. Because workers' compensation costs are such a pervasive concern of company owners and managers, they should desire knowledge of the efforts directed at controlling the related costs. Furthermore, since promoting occupational safety is best facilitated from the top down, company owners and managers should maintain a constant awareness of work-related injuries, as they are indicative of the status of company safety efforts. Because the workers' compensation claims coordinator has the most knowledge of both work-related injuries and the management efforts that have been implemented to control workers' compensation costs, it is fitting that he or she be held responsible for addressing these concerns routinely with the company's hierarchy.

6.4.14 Participating in all safety meetings

Depending on his or her primary job responsibilities, the designated workers' compensation claims coordinator may or may not be responsible for conducting employee safety meetings. However, regardless of who conducts these meetings, the workers' compensation claims coordinator should participate. Not only may the workers' compensation claims coordinator have valuable input for these meetings, such as how past injuries have occurred, but he or she may also gain valuable insight.

6.4.15 Conducting routine evaluations of the company's workers' compensation management program

The value of any undertaking to achieve its intended objectives is determined only through an evaluation of the effort. The policies and procedures that compose a company's workers' compensation management program are no exception. Hence, it is the responsibility of the workers' compensation claims coordinator to evaluate periodically each of the formalized efforts by comparing their intended outcome with the actual result. These periodic evaluations may reveal that policies and procedures are serving their intended purpose or may evidence shortcomings that should be addressed to more adequately fulfill the overall goal of reducing the costs associated with workers' compensation.

6.5 Summary

Many companies mistakenly distribute the responsibilities relating to workers' compensation claims management among several individuals. However, this often leads to a duplication of efforts and a lack of focused leadership and hinders the process of evaluating the effectiveness of those efforts. Because controlling the costs relating to workers' compensation requires effective management, it is necessary to place responsibility for the claims management in the hands of one individual. For the purpose of this text, this individual has been given the descriptive title of workers' compensation claims coordinator.

Because the Workers' Compensation Claims Coordinator is intimately involved in every aspect of workers' compensation, due diligence should be practiced in the selection of the employee assigned to this position. The individual selected for this position should be goal-oriented and empathetic. Additionally, the selected individual should possess well-developed inter-personal communication and organizational skills. Because of his or her knowledge of related facets of the position, the company's safety director or onsite occupational nurse may be an ideal candidate. Regardless of who is selected to assume the duties of workers' compensation claims coordinator, authority commensurate with the magnitude of responsibility must be granted.

The duties of the selected workers' compensation claims coordinator will likely vary among employers. However, it is essential that the workers' com-pensation claims coordinator be involved in every effort that affects effective claims management. Customarily, these duties include coordinating the development and implementation of relevant policies, conducting relevant training, and evaluating the company's efforts to ensure that the intended objectives are being reached. The workers' compensation claims coordinator should also be the central contact person for all work-related injuries, screen claims for evidence of fraud, submit claims to the workers' compensation insurance carrier, and coordinate modified duty assignments. Last, as the workers' compensation claims coordinator is a company's central figure for workers' compensation cost containment efforts, it is essential that this indi-vidual maintain effective communication with all of the parties involved in the claims process.

chapter seven

Understanding and implementing a modified duty program

Contents

7.1   Introduction

Although the most desirable and effective way to reduce workers' compensation costs is to prevent all work-related injuries and illnesses, the reality is that very few companies consistently achieve the goal of "zero injuries." Therefore, it is necessary to develop and implement effective measures to control the costs of injuries once they occur.

One of the most effective measures of managing the costs of workers' compensation claims is modified duty, also known as restricted duty, transitional duty, early return to work, and light duty. Each of these terms reflects an employer's effort to assist employees who suffer a work-related injury to return to gainful employment as soon as possible. It is a temporary position that an employee can fill when he or she is unable to perform regular job duties but can accomplish other productive tasks. Modified duty assignments may be either a modified version of the injured employee's regular job or a completely unrelated task or series of tasks.

Modified duty is an option for employers when, as a result of a work-related injury, an employee's treating physician imposes temporary physical restrictions intended to aid the employee's recovery. It provides employees with a temporary job assignment, within the treating physician's physical restrictions, while keeping the injured employee connected with his or her employer.

Although the term *light duty* was widely used in the past, *modified duty* has gained considerably more acceptance in recent years. The reason for this is simple. Light duty suggests that the work to be performed is effortless or at least requires little physical exertion. As such, the term creates notions that are not necessarily accurate. For many employers, it has become almost an impulse to react to discussion of this topic with the statement, "We don't have any light duty jobs." Conversely, modified duty suggests that an employee's job duties are temporarily modified to accommodate physician-imposed physical restrictions. It is not often that an employer will be heard stating, "We can't modify any jobs here."

This chapter presents the workers' compensation claims management technique of developing and implementing a formal modified duty program. The term *program* implies a set of established policies and procedures directed at achieving a defined organizational goal. The goal of a modified duty program is to return injured workers to gainful employment within the company, as soon as possible following a work-related injury. Thus, a modified duty program is the set of policies and procedures enacted for achieving that goal effectively and consistently.

7.2   Benefits of a modified duty program

A well-developed modified duty program routinely returns employees who have been injured on the job back to work as soon as possible after the injury, while the employee adheres to the physical restrictions imposed by his or her treating physician. Assuming that each employee has been hired to fulfill a need of the employer, the rapid return to work of any employee is a benefit in itself. Nevertheless, a well-designed modified duty program provides a multitude of other benefits for both the employer and the injured worker. Figure 7.1 depicts some of the obvious benefits of a modified duty program, which are discussed in greater detail below.

7.2.1   Benefits to the employer

7.2.1.1   Lower workers' compensation costs

To state that workers' compensation insurance is a significant expense for employers is a tremendous understatement. In fact, next to payroll expenses, workers' compensation premiums for many companies are one of the most costly aspects of doing business. Some companies that have

Employer Benefits

Lower Workers' Compensation Costs

Enables Employer to Ensure Restrictions Are Followed

Sends a Message That Workers' Compensation Is Not a Paid Vacation

Weeds Out Employees Looking for a Free Ride

Employees Return to Their Regular Jobs More Quickly

Employee Benefits

Sends the Message That the Employee Is Valued

Enhances the Employee's Sense of Self-Worth

Eliminates the Psychological Effects of Idleness

Speeds Return to Regular Job

Provides for Rapid Resumption of Salary and Other Interests

Figure 7.1

failed to effectively manage their workers' compensation claims have seen them spiral out of control. Because workers' compensation insurance is such a significant expense, the ability of a modified duty program to help reduce workers' compensation premiums is undoubtedly the primary benefit in the eyes of most employers.

To understand how a modified duty program can affect workers' compensation premiums, one must have a basic understanding of how workers' compensation premiums are affected by past claims. Although overly simplified, it is sufficient to say that a company's workers' compensation premium is increased or decreased according to the workers' compensation claims history for the preceding three-year period, assuming that other factors such as payroll, classifications of employees, and the insurance company rates remain relatively constant. The extent to which the premium is increased or decreased is dependent on the number of claims during that three-year period and the cost (severity) of those claims. This method of premium calculation can be viewed as punishing employers, through higher premiums, who fail to properly manage employee safety and workers' compensation claims and rewarding employers, through lower premiums, who properly manage employee safety and workers' compensation claims.

A detailed explanation of how premiums are affected by past workers' compensation claims is presented in Chapter 5.

Whereas injury-prevention efforts seek to prevent or at least minimize the frequency of injuries, a modified duty program concentrates on reducing the severity of claims after they occur. Specifically, it seeks to eliminate or minimize indemnity (wage-replacement) costs which would otherwise be incurred whenever an employee is off work for an extended period of time as a result of a work-related injury.

As described above, the control of workers' compensation claims costs through the use of modified duty has the systematic and quantitative effect of reducing workers' compensation premiums. However, the implementation of a modified duty program may also have the effect of reducing workers' compensation premiums in a less predictable and structured manner. It is important to note that many workers' compensation insurance carriers allow underwriters a degree of latitude when quoting workers' compensation premiums. This latitude enables underwriters to decrease or increase the premium based on such concerns as anticipated employer cooperation. Hence, the implementation of a modified duty program may "score points" with the insurer and translate into reduced workers' compensation premiums.

7.2.1.2 Enables employers to ensure restrictions are followed

Employees who can be placed in a modified duty assignment have been prohibited from performing specific physical activities. Because physician-imposed physical restrictions are intended to promote timely recovery from injuries, it is in the best interest of employers to ensure that injured workers do not violate these restrictions. By assigning modified duty tasks to employees who are temporarily unable to perform their regular job duties, the employer increases control of the injury recovery process; the supervision of workers assigned to modified duty tasks enables the employer to correct workers who are reluctant to adhere to physician-imposed physical restrictions. Therefore, modified duty enables the employer to ensure that the physician's physical restrictions are being followed, at least while the employee is at work. Conversely, if the employee is off work, the employer has no idea of what the injured worker is doing. While presumed to be "sitting at home," the injured worker could get bored and engage in activities that violate the physician's restrictions. If so, the employer would never know and would have no control over the injury recovery process.

This same concept can be extended beyond the employer's efforts to ensure that employees adhere to the physical limitations imposed by their physician. Many times physicians instruct injured workers to perform stretching and/or strengthening exercises on their own. These exercises are a form of rehabilitative physical therapy and are intended, as are the temporary physical restrictions, to provide for the speedy and complete recovery from injuries. Just as modified duty assignments enable the employer to monitor the activities of injured employees and ensure adherence to physical limitations, modified duty assignments also enable the employer to ensure that injured employees perform physical therapy exercises. This can be done through direct supervision, by having a supervisor or other employee perform the exercises with the injured employee, or simply by designating a time during the work shift at which the injured worker can perform physical therapy exercises with some degree of privacy.

7.2.1.3 Sends message that workers' compensation is not a paid vacation

Modified duty sends a clear message to all employees that a work-related injury does not entitle the injured worker to "paid vacation time." Although the analogy of time off work following a work-related injury and paid vacation following a period of employment longevity may seem to be a stretch, it is worth mentioning that some employees view both as an entitlement.

Unfortunately, some injured workers view time off work as something that they are entitled to receive, regardless of whether or not the time away from work is for the purpose of recovering from an injury. These workers are not necessarily intending to abuse the workers' compensation system. Instead, many have simply been socialized to view time off work following a work-related injury as something that they have earned. To these workers, time away from work following a work-related injury is just as much a benefit as the paid vacation that many companies offer.

By consistently implementing modified duty whenever possible, all employees are made aware that time away from work following a work-related injury is not an entitlement but rather a rare occurrence. In short, by implementing a modified duty program, the expectations of injured workers are shaped to reflect the reality that a work-related injury does not entitle the injured worker to a vacation at the expense of the insurance provider and, ultimately, the employer.

7.2.1.4 Weeds out employees looking for a free ride

Unfortunately there are people in this world who seem to seek a "free ride" at the expense of others and who view workers' compensation as a means to that end. Most employers attempt to avoid hiring that type of person. However, it is difficult to judge a prospective employee's moral and ethical character through a couple of preemployment interviews and reference verifications. This situation is further complicated by the fact that past employers are apprehensive about making negative comments and by laws that limit questions that are permissible to ask an applicant. Inevitably, a few people of questionable moral fiber slip through the cracks and become employees as well as thorns in the sides of employers.

A modified duty program can help "weed out" employees who want to abuse the workers' compensation system for personal gain. How a modified duty program accomplishes this feat is quite simple. Whenever possible, a modified duty assignment should be provided to any employee who has sustained a work-related injury that prohibits performance of his or her regular job duties but has been released by the treating physician to perform work with specific temporary physical restrictions. In such situations, modified duty is the litmus test to determine if the injured employee is inclined to cooperate with the employer or to abuse the workers' compensation system. If the employee returns to work and performs the modified duty assignment, the employee is not likely a person prone to seek a "free ride." However, it is not uncommon for an employee to refuse a modified duty assignment, even if it is work that he or she is physically capable of performing. Such a refusal may indicate a general unwillingness to cooperate or may even reveal a person who views workers' compensation as a desirable alternative to conventional employment. As a result of a refusal to return to work within the context of modified duty, the insurance provider will likely cease indemnity benefits to the injured employee, and it may constitute justification for the termination of employment.

7.2.1.5   Employees return to their regular jobs more quickly
Modified duty has the goal and result of returning the injured employee to some level of productive work more quickly. However, many employers do not realize that allowing an employee to return to work in a modified duty status generally results in the employee returning to his or her regular duties, at a normal level of productivity, more quickly. This suggests that modified duty actually enhances the rehabilitation of injured employees. Conversely, an employee who is not provided the option to return to work within the limitations of the physician's temporary physical restrictions is more likely to malinger or to require greater time to reach pre-injury levels of productivity.

7.2.2   Benefits to the employee

7.2.2.1   Sends the message that the employee is valued
Although income and other tangible benefits are important to all workers, so are the intangible benefits, such as the feeling of being of value to the company. Because most people work to provide themselves and their families with a comfortable living, most would hold that salary, medical benefits, retirement plans, etc. are their primary concerns. However, when an employee is injured on the job and is faced with potential time away from work, anxiety can take over. The employee's primary concern may then switch to thoughts of how he or she is perceived by the company. "Does the company view me as a problem, or am I still a valued employee?" and "Do I still have a future at this company?" may be questions that an injured employee may ask himself or herself.

Employees' perceptions of their value to the company largely depend on how the company responds to them following an injury. When a worker is injured on the job and is released to return to work with temporary physical restrictions, the employer has two choices. The employer can either modify the worker's duties and allow the employee to return to work within the physician-imposed physical restrictions, or the employer can tell the worker that he or she may not return until completely released from the physician's care (with no physical restrictions).

These two different responses not only indicate the employer's view toward modified duty but also convey a message to the injured worker. To modify workers' job duties to allow a return to work sends the message that workers are valued not only when they are 100% able but also when less than 100% able. To refuse to implement modified duty whenever possible sends the message that employees are of limited value to the company.

7.2.2.2   Enhances the employee's sense of self-worth

When workers feel needed and valued by employers, it increases their sense of self-worth. However, regardless of whether they feel valued by the employer, most workers identify themselves with their jobs. If in doubt, ask several people to describe themselves. Few will be able to utter a few sentences before mentioning their job or their employer. Similarly, many individuals of retirement age continue to work, even if it is not a financial necessity. As such it is only logical to correlate an individual's sense of self-worth with his employment status and job position.

With this correlation in mind, it is easy to see that an employee's sense of self-worth decreases while the employee is off work and will similarly increase when the employee returns to work and resumes normal levels of productivity. Through a modified duty program, employees regain their connection with their employer more quickly. Although an employee temporarily may not maintain the same levels of productivity as compared with pre-injury levels, he or she maintains the sense of belonging and identity that employment provides.

7.2.2.3   Eliminates the psychological effects of idleness

Anyone who has taken a two-week vacation from their job, or who has been away from work for an extended period due to an illness or other reason, knows that it is tough to get "back in the swing of things" up on returning. Whereas time away from work often provides a necessary break, the work routine is disrupted by an extended absence. On the first day back to work after a extended period away, it may even be difficult to convince oneself to get out of bed and go to work, not to mention jumping right back into the work routine with both feet.

The same phenomenon applies to employees who are absent from work due to a work-related injury. Their absence from work creates a psychological hump that they must cross when returning to work. However, for employees who are absent from work for an extended period due to a work-related injury, the psychological hump is compounded by other factors. Beyond the impact that being away from work has, the injured worker may have very real concerns and fears of how he or she will be viewed by the employer and supervisors, the reactions of coworkers, and how (if at all) he or she will be able to perform the job task that caused the injury. Due to idle time while away from work, these concerns and fears are constant thoughts that multiply in intensity the longer that the employee is away from work.

By allowing the worker to return to work through a modified duty program, the employer enables the worker to forego the need to overcome the psychological hump of returning to work after an extended absence. In many circumstances, an employee can be permitted to return to work the day following an injury as long as the physical restrictions imposed by the treating physician are followed. Even a modified duty assignment involving only a few hours of work per day helps prevent the need to overcome the psychological hump, by keeping the injured worker connected to his employer and coworkers.

7.2.2.4   Speeds return to regular job
Through modified duty, a worker who has been injured on the job is enabled to return to employment even though his treating physician has imposed temporary physical restrictions. Because of the temporary physical restrictions, in many circumstances the injured worker is provided with modified duty completely different from the regular job duties. For example, an employee whose regular job involves driving a truck may be provided with a modified duty assignment involving clerical work. However, one of the beneficial attributes of modified duty is that the worker is likely to return to the regular job more quickly than if he or she had not been provided the modified duty assignment. Both employees and employers benefit from the timely return of the injured worker to his or her regular job duties.

As mentioned previously, our sense of identity is, at least partly, tied to what we do for a living. In the above example, the injured worker is a truck driver. This employee does not identify himself or herself as a secretary, receptionist, or other clerical worker. As such, the injured employee is motivated to return to his or her regular job, and identity, more quickly. Of course, the injured worker must adhere to his or her treating physician's physical restrictions and cannot return to regular duties until released to do so. Nevertheless, the desire to return to regular job duties and regain his or her identity may encourage the employee to be more active in his or her recovery.

7.2.2.5   Provides for rapid resumption of salary and other benifits
When an employee is off work due to an injury or illness that is compensable under state workers' compensation legislation, the workers' compensation provider typically pays the employee in lieu of regular wages. The amounts paid are based on state laws and are generally less than the employee's normal wages (typically two-thirds of the employee's normal wages). Conversely, when an employee returns to work on a modified duty status, his employer pays him. As such, return to work, even on a modified duty status, generally results in the employee's resuming his or her regular wage.

Enabling an employee to return to work on modified duty status also enables the employee to continue with uninterrupted seniority and progression. Similarly, deductions for such items as profit sharing and insurance are continued while performing work for the employer, whereas these deductions are generally not continued when an employee is receiving workers' compensation indemnity benefits.

7.3   Resistance to implementing modified duty

Despite the numerous benefits of modified duty, resistance to its implementation is far from uncommon. This resistance can come from almost anyone involved with workers' compensation claims, including the injured employee, the treating physicians, or the employer, generally represented by a supervisor or manager. The workers' compensation insurance company (or third-party administrator) will rarely, if ever, offer resistance to an employer desiring to provide an injured employee with a modified duty assignment, assuming that it is permitted by the employee's treating physician. However, the other parties to a workers' compensation claim are sometimes reluctant to embrace the concept of modified duty.

Those who resist the use of modified duty almost always view their position as completely justifiable. They view their stance as a reason or justification for which modified duty is not feasible in their situation or in their business. In reality, the *reasons* presented for why they *cannot* implement modified duty are more accurately dubbed *excuses* for why they *will not* implement it. Although there are some situations where modified duty is simply not feasible, many individuals who initially resist modified duty are convinced of its inherent benefits once they become familiar with the flaws of their objections.

7.3.1   Resistance from employers

Perhaps the greatest resistance to implementing a modified duty program comes from employers. This is surprising, as employers stand to benefit the most from its use. It is likely that much employer resistance to modified duty is a result of not being familiar with how its use can benefit the company. Additionally, many owners, managers, and supervisors in well-established companies who have refrained from the use of modified duty in the past may be apprehensive of change.

Some of the most common excuses for resisting the implementation of a modified duty program are presented below, accompanied by ways to refute them. In addition to those reasons, in any discussion of modified duty it is important to note that failure to implement modified duty whenever possible negates all of the benefits presented earlier in this chapter.

7.3.1.1   "We don't have light duty jobs"

When faced with the opportunity to provide an injured employee with a modified duty assignment or to implement a formal modified duty program, many employers are quick to resist by stating their company does not have any "light duty" jobs. As previously mentioned, the frequent use of this excuse is perhaps the primary reason that there has been a concerted effort to refrain from using the term *light duty*. By replacing it with *modified duty*, employers who are reluctant to accept the concept are forced to rethink their negative response. Hence, a reluctant employer would be forced to say something like, "We are not willing to modify an employee's duties." This latter response is probably a more accurate sentiment, as it identifies unwillingness rather than an inability to implement a modified duty program.

With rare exceptions for companies with few employees and whose employees perform a very limited number of tasks, all employers should be able to modify an employee's duties temporarily to accommodate physician-imposed temporary physical restrictions arising from a work-related injury. The stumbling block that most employers have is not the inability to implement modified duty, nor are they necessarily uncooperative and stubborn. Rather, a preconceived and limited view of modified duty is the biggest obstacle which must be overcome.

Employers with a narrow view of modified duty often assume that it can be implemented only if the company has an existing full-time position that would not violate the injured worker's physical restrictions. Furthermore, they may also erroneously assume that if such a job exists, it must be one that is not currently staffed by another employee and that does not need to be altered to accommodate the injured worker. Similarly, many employers who do not fully understand modified duty believe that it is synonymous with clerical work. With these narrow views prevailing, it is easy to see why some employers resist it by saying, "We don't have any light duty jobs."

To overcome the obstacle of a limited view of modified duty jobs, employers must understand what modified duty encompasses. It is not limited to allowing an injured employee to perform temporarily an existing job that is not currently staffed, nor is it limited to clerical positions. Instead, it can mean that the employee returns to his or her regular job and performs those tasks that he or she is still able to perform, while other employees perform the other tasks. Similarly, the regular job of the injured employee can be modified to enable that employee to perform some or all of the tasks of his or her regular job. Other optionsinclude allowing the injured employee to perform tasks completely unrelated to his or her regular job duties or to identify tasks within the job descriptions of a number of employees which the injured employee is able to perform, and to charge the injured employee with performing those tasks. Those employees can then concentrate their efforts elsewhere or be rewarded with a temporarily lighter workload.

Various attributes of the job may be altered to facilitate an employee's early return to work. These include, but are not limited to, the manner in which regular job duties are performed, the tasks that the employee performs, the length of time that an employee works, and the speed or productivity rate at which the employee works. For now, it will suffice to say that the ability to implement modified duty is not so much a question of "if" a company is able but rather of how creative an employer is willing to be in order to facilitate the early return to work of an injured employee and thereby reap the numerous benefits of doing so.

7.3.1.2 The whole man or no man stance

When given the opportunity to facilitate early return to work through modified duty, many employers indicate that they want the injured employee to be at 100% capacity before he or she returns to work. With the exception of employers who claim to have no "light duty" jobs, this excuse is perhaps the second most common for not providing modified duty assignments. In fact, this response is so common among employers who are reluctant to use modified duty that it has been tagged the "Whole Man or No Man Stance terminology."

Employers with this attitude are merely stating that they have hired an individual to perform a specific job, and if that person is unable to perform that job at 100% effectiveness, he or she is of no use to the company. Despite this rigid stance, few (if any) employers could honestly declare that all of their employees give 100% every day.

In addressing the Whole Man or No Man Stance, employers must ask themselves if they would rather pay an employee to be 60% productive or pay an employee who is 0% productive. The question is that simple. Because the cost of past workers' compensation claims is considered when calculating future premiums, the employer is likely to incur an increased workers' compensation premium (or at least a decrease in the rate of premium reduction) if the employer does not embrace modified duty. Therefore, it can be reasoned that the employer ultimately pays for the employee's wages when the employee is off work as a result of a work-related injury. Through providing a modified duty assignment, the employer pays the employee a wage during his or her period of recovery and by doing so receives some level of productivity from that employee. Conversely, an employer who adheres to the "Whole Man or No Man Stance" receives 0% productivity from that employee, thereby incurring the financial burden for that decision through increased future workers' compensation premiums. To add insult to injury, the employer who refuses to bring an employee back to work until he or she is 100% able must also incur the additional expenses associated with having another person perform the injured employee's regular job duties and should also take into consideration that employees who return to work in a modified duty assignment are likely to return to their regular job duties more quickly than employees who remain off work during their recovery period.

Beyond the issue of cost and productivity, the Whole Man or No Man Stance sends the message to all of the employees that they are of no value to the company if they are not at 100% effectiveness. Since none of us are 100% effective every day, employers should consider the impact that a policy such as this has on the morale and attitudes of employees and their perception of management. In a time when qualified employees are difficult to find, and even more difficult to retain, a policy that does anything but strengthen employee/employer relations is ill-advised.

Not only does the Whole Man or No Man Stance result in an employer paying for 0% productivity and contribute to poor employee/employer relations, but it may also create a serious legal problem. It is important to consider the implications if the injured employee never reaches what the employer considers to be 100% effectiveness. In such situations, the injured employee then has a legal disability, regardless of whether it is an actual physical disability or an employer-perceived disability. If reasonable accommodations can be made to facilitate the employee's return to work, failure to accommodate the employee and to return the employee to work may constitute a violation of the Americans with Disabilities Act (ADA), the financial implications of which far exceed normal workers' compensation claims costs. In addition to the financial penalties, a ruling against the employer may still require the company to return the employee to work.

7.3.1.3 *"I don't want to be sued if the employee re-injures himself"*

Some employers resist implementing modified duty for fear that they will be sued if an employee re-injures himself or herself while at work before being completely released from the treating physician's care. In a society that has become so litigious that a customer has successfully sued a restaurant for burning herself with the hot coffee that she was served, the fear of being sued for virtually anything is not without merit. However, an employer need not be concerned with litigation from implementing a modified duty program, as long as the employer does not require an employee to perform tasks that are obviously in violation of the physician-imposed physical restrictions. Furthermore, it is important to remember that workers' compensation is not only a protection for workers but a protection for employers — in the absence of negligence, it is the sole remedy for compensation from work-related injuries. As such, it is a generally true statement that the only employees who can successfully bring a civil suit against an employer for a work-related injury are those who are not covered under a worker's compensation policy or those who can prove employer negligence as the causative factor of an injury.

To further refute this excuse for failing to implement a modified duty program, employers should consider one of the previously described benefits of such a program — the employer has control over what an employee does at work but no control over what the employee does away from work. Therefore, to the extent that the employee's activities are under the control of the employer, the employer can help prevent the aggravation of the employee's injury through ensuring strict adherence to the temporary, physician-imposed physical restrictions. In fact, as previously stated, the use of modified duty may have a therapeutic effect, thereby providing for more rapid recovery from an injury.

7.3.1.4   It's too much trouble to modify work assignments

Some employers claim that modifying the work assignment of the injured employee (and perhaps the work assignments of other employees) is too much trouble. It is true that developing modified duty assignments for employees who sustain work-related injuries is not always an effortless task. However, employers must weigh the pros and cons of implementing modified duty and ask if the effort to implement modified duty is so great that they are willing to forgo the benefits.

To implement modified duty, the injured employee might be charged with performing a task that needs to be performed but for some reason has not been performed by another employee. This may include such things as painting, cleaning, or filing which has not been done because of time or manpower constraints. A modified duty assignment might also involve having the injured employee perform an existing job in which there is a current vacancy. In both of the above situations, the implementation of modified duty is relatively effortless, assuming that the task can be performed within the physician-imposed temporary physical restrictions.

Although some situations, such as those mentioned above, lend themselves to the implementation of modified duty without much effort or planning, more often than not such ready-made modified duty positions are not readily available. Thus, providing an employee with a modified duty assignment frequently involves greater effort. Commonly the use of modified duty involves having the injured employee perform his or her regular job duties which have been modified to accommodate the physician-imposed physical restrictions. It is also common to have the injured employee perform various tasks within the job descriptions of other employees. With either of these two applications of modified duty, not only may it be a demanding chore to identify specific tasks for the injured employee to perform, but other employees are also affected by adjusting their work activities to complement or compensate for the modified duties of the injured employee.

Although identifying specific tasks that an injured employee is able to perform may be a demanding task, it is simplified through preparedness. Consider the analogy of a company responding to an impending tornado. If the employer has not prepared, a last-minute attempt to identify a safe sheltering area, to alert employees, and to get employees into that area in an orderly manner will likely cause confusion, disorder, and chaos. Conversely, an employer who has a plan of action for responding to tornado alerts will likely implement the plan with relative ease. Similarly, an employer who establishes a modified duty program before there is a need to provide an employee with a specific modified duty assignment will be able to facilitate an employee's early return to work with relative ease.

As described in section 7.4, a well-prepared modified duty program will incorporate written job descriptions for existing jobs, a list of past modified duty assignments, and a list of potential modified duty assignments for common physical restrictions. These elements, along with defined responsibilities relative to modified duty, will make the process of providing an employee with a modified duty assignment considerably easier.

7.3.1.5   It will create morale problems with other employees

Some employers are concerned that allowing an employee to return to work following a work-related injury on a modified duty status will create morale problems with other employees. Employers may fear that the other employees will feel that the injured employee is getting a lighter workload as a result of his or her injury, and that an extra burden is being placed on them to compensate. In short, employers may feel that other employees will view modified duty as a reward program for employees who have been injured, at the expense of the employees who have not.

Whereas this concern is probably warranted in many situations, it is illogical to consider the use of modified duty as creating a potential morale problem without considering the alternative. To the extent that modified duty appears to other employees as a reward program for employees who have been injured on the job, the alternative of remaining off work should be scrutinized even more closely.

If an employee is able to return to work with temporary, physician-imposed physical restrictions but is not provided with a modified duty assignment, he or she is remanded to remain off work until released to return without physical restrictions. In such circumstances, the employee, although able to perform some degree of work, is permitted to remain off work and collect wage-replacement benefits for doing so. While the employee is off work, other employees are forced to compensate for the injured employee. Hence, viewing the two options logically, the conclusion is that the morale of other employees would be more negatively affected by a situation in which the injured employee is permitted to remain off work and contribute nothing to the efforts of the company than if the injured employee returned to work through a modified duty program and performed at some level of productivity.

7.3.2   Resistance from employees

Employers are not the only people who resist the concept of modified duty. Sometimes the injured employee is opposed to modified duty as well. An employee's resistance to accepting a modified duty assignment generally reveals itself in one of two underlying excuses. One is that the employee feels that he or she is entitled to time away from work, as well as workers' compensation indemnity benefits, as a result of sustaining a work-related injury. The other is that the injured employee feels that the modified duty assignment involves a task that he or she does not want to perform.

Several things can be done to increase employee acceptance of modified duty, including:

- Promoting it as an employee benefit rather than a workers' compensation claims management strategy,
- Explaining to employees how modified duty can benefit them, and
- Considering the input of the injured employee when developing a suitable modified duty assignment.

Although a modified duty program works best when it is accepted and viewed as a beneficial program by both employer and employees, its implementation does not necessarily require the embrace of the injured employees for whom modified duty assignments are provided. Generally speaking, an employee has no more right to refuse a modified duty assignment than he or she has to refuse to abide by written safety rules. Many companies view the refusal of a modified duty assignment as a voluntary termination of employment. Regardless of employment status, the refusal of modified duty almost always results in a termination of indemnity benefits.

7.3.3   Resistance from physicians

Depending on experience with treating occupational injuries, a physician may be reluctant to release an injured employee to return to work with temporary physical restrictions. This reluctance may be based on the physician's professional medical opinion, unfamiliarity with the employer's operations, or fear of medical malpractice litigation.

The treating physician may believe that an injured employee would recover from his or her injury more quickly if he or she is not involved in any type of work (whether physically demanding or not). Whereas individuals who lack the professional medical training cannot expect to successfully challenge a physician's prescribed treatment of an injury, the rehabilitative benefits of early return to work are documented in many studies.

Some states permit employers to designate a single physician or list of physicians from whom employees must seek treatment for work-related injuries. In states in which this practice is permitted, one of the primary concerns when selecting physicians is their willingness to prescribe temporary physical restrictions as opposed to time away from work. Furthermore, once selected, physicians should become very familiar with the tasks that employees perform, as well as the company's modified duty program, so they can more accurately assess the employees' ability to return to regular job duties.

In states that permit the employee to select his or her treating physician, there is little that can be done, other than obtaining a second opinion, about physicians who routinely prescribe time off work when it is not necessarily warranted. However, describing the company's modified duty program to the treating physician may influence the physician enough to prescribe modified duty restrictions with greater frequency.

7.4 Getting started

Although a company can employ modified duty with little or no planning, the process is made much simpler if preparatory measures are taken. These measures include creating functional job descriptions, identifying potential modified duty assignments, developing a written modified duty program, communicating with medical providers, and conducting employee training.

7.4.1 Creating functional job descriptions

Often an employee who has been released to return to work in a modified duty capacity is able to perform the essential tasks of another job within the company. Other times the injured employee may not be able to perform any complete job within the company but may be able to perform a specific task that is a part of a job, or a series of tasks that are parts of several jobs. If a company has planned ahead and assessed the physical requirements of the various tasks performed in the course of business, then the process of matching the injured employee to a suitable modified duty assignment is made quickly and easy.

It is the responsibility of managers, supervisors, team leaders, and all other employees who serve in a supervisory capacity to develop functional job descriptions — supervisory personnel have firsthand knowledge of what each job entails within their respective areas of responsibility. Whereas the format that is used to document the physical requirements of each job or task may vary from one company to another, there are several guidelines that should be followed.

Before any job can be assessed to determine the essential physical requirements, all of the tasks that compose the job must be enumerated. To do this it is necessary to understand the difference between "jobs" and "tasks" and when the physical requirements of each task should be assessed separately vs. assessing the physical requirements of an entire job. Whereas "secretary" is a job, it comprises a number of different tasks, such as answering the telephone, filing, and typing. Because each of these tasks is independent of the others, an employee assigned to a modified duty position may be able to perform one or more of these tasks without performing all the tasks of a secretary. Therefore, a separate assessment should be performed for each task. Likewise, "truck driver" is a job that comprises the tasks of driving, conducting pre-trip inspections, and loading and unloading freight. If these tasks cannot be performed independently (by different people) then the tasks are interdependent. In such a situation, if an employee assigned to a modified duty position is not able to perform one of the tasks, then he or she is unable to perform them all. In such situations a single assessment of all interdependent tasks of a job should be performed as if they were a single task.

Second, regardless of the format used to document the essential physical requirements of the jobs within a company, all supervisors should use the same format. The workers' compensation claims coordinator can then easily catalog the data. Figure 7.2 illustrates a simple form that can be used to document the physical requirements of each task performed in the course of business.

Functional Job Description

Company Name	
Department	
Job Title	
Description of Task	
Frequency of Task	Routine / Non-Routine

Has this task been used as a modified duty assignment in the past? *(If yes, indicate rate of pay)*

Physical Activity	Description
Lifting Over 10 lbs.	
Lifting Over 25 lbs.	
Lifting Over 50 lbs.	
Lifting Over 75 lbs.	
Lifting Over Shoulder Level	
Squatting	
Carrying	
Standing	
Sitting	
Bending	
Climbing	
Walking	
Twisting	
Repetitive Motion	

Figure 7.2

7.4.2   Identifying potential modified duty assignments

Whereas assessing the physical requirements of tasks performed in the normal course of business is a process from which potential modified duty assignments can be identified at a later time, it is likely that supervisors, managers, team leaders, and other personnel serving in a supervisory capacity can identify potential modified duty assignments that can be performed within their respective departments. One of the most challenging aspects associated with modified duty for supervisors is being "put on the spot" to identify a modified duty task quickly in response to an employee who is ready to return to work on modified duty status. However, supervisors are often able to identify potential modified duty tasks when given several weeks to consider all the options. By documenting these potential modified duty assignments, supervisors make the implementation of modified duty a virtually effortless chore.

When identifying potential modified duty assignments, supervisors do not have the benefit of knowing what specific physical restrictions are going to be imposed. For this reason, it is beneficial to consider those physical restrictions most commonly associated with modified duty. These include lifting restrictions, limited use of one arm or hand, limited use of one leg or foot, and limited standing.

When considering potential modified duty assignments, supervisors should realize that the identification of potential modified duty assignments is limited only by their creativity. The potential modified duty assignments may be tasks currently performed in the course of business and can be tasks with or without modifications to the "normal" manner in which they are performed. They may also be tasks which are performed periodically, such as preventive maintenance, painting, and organizing stock. Furthermore, potential modified duty assignments may be identified by considering tasks that have been postponed due to lack of time, such as filing, updating records, and conducting training. Another option for identifying potential modified duty assignments is to consider how an employee with temporary physical restrictions could benefit the company's safety efforts. This may include conducting safety meetings or safety inspections, assisting in the development of new safety programs, or researching safety literature to identify additional safety efforts that may benefit the company.

Because each potential modified duty assignment identified has unique physical requirements, each should be assessed in the same manner that job tasks are assessed. This provides a consistent manner in which to review documentation when attempting to pair a specific employee with a modified duty assignment. As such, the same form that is used for assessing the physical requirements of tasks performed in the normal course of business can also be used to document the physical requirements of potential modified duty assignments.

7.4.3 Communicating with medical providers

Since modified duty is not an option if an employee's treating physician imposes time away from work as opposed to allowing the employee to return to work with temporary physical restrictions, it is essential to gain the cooperation of the medical community. Occasionally, the injured employee misleads the treating physician into believing that the company is unable or unwilling to accommodate temporary physical restrictions. In such situations the treating physician may impose time away from work as the only medically feasible option. Thus, a primary goal when preparing to implement a modified duty program is to inform likely medical providers of the company's willingness to provide work for injured employees who are temporarily unable to perform their regular job duties because of a work-related injury.

Medical providers are not only reluctant to release an employee to return to work with temporary physical restrictions when uncertain if the company is willing or able to accommodate the restriction, but they are also reluctant to permit modified duty assignments when the injured employee exaggerates the physical demands of his or her job. Thus, it is not only beneficial to inform potential medical providers of the company's modified duty program but also to familiarize them with the work performed by employees and the physical demands of those jobs. This familiarization can be accomplished through either providing functional job descriptions or encouraging the physician to visit the workplace and observe the work that employees perform.

7.4.4 Conducting employee training

In addition to communicating with the medical providers who are likely to treat injured employees, it is necessary to conduct training for the employees who, if injured in the course of employment, would be affected by the modified duty program. Since it cannot be determined which employees will be injured, this training should be provided to all employees. Training employees about a modified duty program, however, need not be as intensive and time-consuming as safety-related training. Instead, this training should simply facilitate the education of all employees, supervisors, managers, and executives of the content of the program and their roles and responsibilities relative to the program. The goal should be to have each employee know that modified duty will be used whenever applicable, that it is an employer-provided benefit, and that success of the program is dependent on individual employees performing their assigned responsibilities.

Since the workers' compensation claims coordinator will be the individual within the company who has the most knowledge of the company's modified duty program, that individual would be the most appropriate person to conduct this training. As such, the workers' compensation claims coordinator must be familiar with not only his or her role but also those of all other employees as well. However, it is notable that most workers' compensation carriers will provide someone to assist in conducting this training on request.

7.4.5   Developing a written modified duty program

Many companies have written policies and procedures that address everything from sexual harassment to employee drug-testing. Additionally, government agencies, such as OSHA, require almost all employers to develop and implement specific written safety programs. These documents are intended to fulfill the broad objective of protecting the company and its employees. A written modified duty program is certainly no exception, as it is intended to protect both the employer and the employee by promoting the benefits discussed earlier in this chapter.

7.4.5.1   Reasons for developing a written modified duty program

A written modified duty program provides a clear definition of the company's position relative to modified duty and the underlying reasons for adopting the program. It identifies when modified duty is applicable and when it is not. Additionally, having the modified duty program in writing provides clear descriptions of individual responsibilities and identifies a specific series of events that should be followed. In short, by defining the company's modified duty program in the form of a written document, employees know what to expect from the employer as well as what is expected of them.

In addition to defining the company's policies relative to modified duty, a written modified duty program is evidence of a company's commitment to the concept of modified duty. This may mean different things to different groups. To a workers' compensation carrier, it may mean that the employer is willing to do its part to manage workers' compensation claims. To employees, a written modified duty program may be interpreted as evidence of the employer's commitment to provide employees with a steady paycheck, even when an employee cannot perform at 100% capacity. For physicians, a written modified duty program might imply that the company is willing to work cooperatively within the physical abilities and restrictions of the injured worker to provide for optimal recovery. Workers' compensation carriers are so eager for this evidence of management commitment to modified duty that a written modified duty program, combined with an effective safety program, may result in a moderate reduction of workers' compensation premiums. Similarly, physicians who are convinced of a company's commitment to modified duty are more likely to prescribe modified duty as opposed to time off work.

Because a written modified duty program defines the roles and responsibilities of all of the entities involved, it enables the program to be administered in a consistent manner. Hence, equal effort is expended to providing a modified duty assignment to each employee for whom a modified duty assignment is applicable. Furthermore, particulars such as the method of determining the rate of pay that an employee receives and the maximum length of a modified duty assignment remain consistent for all employees who are provided with a modified duty assignment. Conversely, the absence of written procedural guidelines creates the likelihood of inequitable applications of the policy, which may lead to morale problems or even formal complaints regarding equal treatment.

Although a written modified duty program is essential to ensuring the consistent application of the modified duty in the absence of personnel changes, it is even more essential when considering the reality of employee turnover. Whether turnover involves the company's workers' compensation claims coordinator, a supervisor, or any other employee, changes in personnel result in workers being introduced to new responsibilities. Without a written modified duty program, new employees and employees who have been promoted within the company do not have definitive information relative to the company's stance on modified duty, thereby virtually ensuring the inconsistent application of modified duty.

The adage "failing to prepare is preparing to fail" is particularly true with respect to the development of a modified duty program. Since the goal of a modified duty program is to return injured workers to some form of productive employment within the company as soon as possible after being released to do so by their treating physician, anything that can be done to reduce the amount of time required to achieve that goal is beneficial. For this reason, it is essential to prepare for the use of modified duty by identifying the responsibilities of the individuals involved and the procedures that must be followed for the unencumbered application of the program. Without this preparation, a company is forced to be reactive rather than proactive. As such, failing to develop a written modified duty program means that identifying what should be done and who should do it consumes the time that should be used for enacting the policies of a previously prepared modified duty program. Furthermore, without the established procedural guidelines provided by a written modified duty program, the same process of identifying what should be done and who should do it must be performed each time that an employee is eligible for modified duty.

In addition to the time and effort that is saved by having previously developed written procedural guidelines and assigned responsibilities, the documentation included in a modified duty program can also conserve time and effort. Beyond the document itself, a modified duty program should include provisions for documentation relating to the program. Probably the most significant documentation, in terms of saving time and effort, consists of the records indicating the physical requirements of various job tasks, the documentation of potential modified duty assignments, and the documentation of modified duty assignments that have successfully been implemented in the past. When faced with identifying a modified duty assignment for an employee, these documents can save a considerable amount of time that would otherwise be spent assessing the entire workplace to identify tasks that the injured person can perform.

7.4.5.2   Elements of a written modified duty program

For as many companies that have written programs, there are likely an equal number of different styles and formats. Whereas the programs written by some companies are lengthy and detailed, other companies have written programs that are rather brief and broadly defined. There are benefits and drawbacks to both approaches. Although a lengthy written program provides detailed instructions and little room for varied interpretation, it can be so lengthy that few employees actually read it. Conversely, a short written program may be read by more employees but lack sufficient detail to address adequately specific procedures or individual responsibilities. Regardless of the format or length, a modified duty program should be concise but with sufficient detail. It is also essential to write the program in a manner that can be readily comprehended by its intended audience.

Furthermore, because the written program will be used as a reference, it should be designed for that purpose. The content of the program should be divided into segments with descriptive headings. This text describes the likely content of a modified duty program, using the general headings of "Purpose," "Scope," " Management Commitment," "Employee Involvement," "Program Administration," "Education and Training," and "Program Evaluation." With the program divided into manageable segments such as these, employees are able to quickly locate specific information within the program. A sample modified duty policy is provided in Appendix D.

7.4.5.2.1   Purpose.   This section provides a synopsis of the program and identifies the impetus for its implementation. In addition to the primary purpose for implementing the program, there are likely to be collateral objectives. These objectives or goals should be enumerated in this section. This is essential, as it not only provides employees with a better understanding of the intent of the program but also provides criteria by which the program can be evaluated to determine its effectiveness.

7.4.5.2.2   Scope.   This section identifies the employees to whom this program is applicable. Furthermore, this section identifies the specific conditions under which modified duty assignments are applicable. For example, because it has no bearing on workers' compensation costs, a company may choose to disallow the implementation of modified duty assignments for situations in which the injury sustained by an employee is not work-related.

7.4.5.2.3   Management commitment.   This section identifies what the employer is contributing to the program, such as establishing written procedures, assigning a workers' compensation claims coordinator, conducting employee training, maintaining relevant documentation, and conducting periodic evaluations of the modified duty program. These activities are evidence to both employees and to outside entities that the employer is committed to the success of the program.

On a more practical note, this section of the modified duty program also identifies the roles and responsibilities of management personnel for fulfilling the objectives of the program. It itemizes the responsibilities of the workers' compensation claims coordinator relative to the modified duty program.

7.4.5.2.4 Employee involvement. This section identifies the responsibilities of employees, including supervisory staff, relative to modified duty.

7.4.5.2.5 Program administration. This section delineates specific policies within the overall modified duty program. Included in these policies should be an explanation of the amount of pay that an employee will receive while performing a modified duty assignment as well as an explanation of the implications of an employee's failure to accept a modified duty assignment. The maximum allowable length of a modified duty assignment can also be explained. Furthermore, this section may include a brief description of the documentation that is maintained in conjunction with the modified duty program.

7.4.5.2.6 Education and training. Employees must receive training for them to know what to expect from the company and what is expected of them as employees. This alleviates the anxiety and uncertainty so often associated with workers' compensation claims. This section of the modified duty program should thus identify when employees will be trained, such as during orientation, and should provide an outline of the training content.

7.4.5.2.7 Program evaluation. The success and value of any program is determined only by an evaluation of that program. This section identifies the means of evaluating the program's effectiveness as well as a schedule for completing the evaluation.

7.5   The modified duty process in action

Once the modified duty program has been developed, physical requirements of job tasks have been documented, potential modified duty assignments have been identified, medical providers have been familiarized with the employees' job duties and the company's modified duty program, and the employees have received training, then the process of implementing modified duty whenever it is applicable is rather simple. Although specific attributes of a company's modified duty program may differ slightly from those that are described below, the basic concepts should be consistent.

7.5.1   Accompany the injured employee to the medical provider

Regardless of whether medical treatment is an emergency or nonemergency, and regardless of whether the employee's treating physician is a member of a managed care program or the employee's family physician, it is a good practice for the injured employee's immediate supervisor to accompany the injured employee to the medical treatment provider. This procedure not only sends the message to the injured employee that he is valued but also enables a company representative to explain the company's modified duty program to the treating physician and to express the company's willingness to accommodate physical restrictions, if imposed. Often physicians are much less reluctant to prescribe temporary physical restrictions, as opposed to time away from work, when they are aware of the employer's willingness to accommodate the restrictions and to participate in the injured employee's recovery. Furthermore, by accompanying the employee, the supervisor is able to provide the treating physician with an unbiased description of the employee's job duties, both verbally and in the form of written functional job descriptions. Note that it is not uncommon for an employee to exaggerate the physical demands of his or her job in an attempt to persuade the physician to prescribe time off work as opposed to modified duty.

7.5.2   Obtain physical restriction documentation

Once the modified duty program is implemented, the workers' compensation claims coordinator is responsible for seeking clarification of any vague physical restrictions. To be useful, physician-imposed physical restrictions must be very specific and should be documented. This documentation should indicate specific physical restrictions (such as "no lifting over 50 pounds") and should specify the period for which the restrictions are valid (such as "10 days"). If the physician-imposed physical restrictions are not specific, the workers' compensation claims coordinator should contact the physician immediately and request the physician to provide clarification in writing.

When an injured employee's treating physician provides a treatment plan that involves time off work, as opposed to allowing the employee to return to work with temporary physical restrictions, the workers' compensation claims coordinator should contact the treating physician, explain the company's modified duty program, and stress that the company is willing to accommodate temporary physical restrictions during the employee's recovery period. This discussion is often sufficient reason for the treating physician to rethink the initial decision. This is one area in which professionalism and good interpersonal communication skills may prove effective. However, the physician will not always permit the employee to return to work during the recovery period, and a modified duty assignment should not be contemplated if the treating physician has not provided written temporary physical restrictions.

7.5.3   Conduct a post-treatment meeting

As soon as possible after the initial treatment of a work-related injury, the employee should return to the company and meet together with his or her immediate supervisor and the company's workers' compensation claims coordinator. The primary purpose of this meeting is to arrange a modified duty assignment and a schedule for the implementation of that assignment.

In circumstances in which the employee is released to return to work with no physical restrictions, and in circumstances in which the employee is instructed to remain off work for a recovery period, a meeting between the employee, his or her supervisor, and the workers' compensation claims coordinator would not be necessary for the purpose of discussing modified duty, but it may be beneficial so all parties are aware of the details of the loss event and the outcome of the initial medical treatment.

The supervisor's involvement in these meetings is important. Although the previously accumulated information relative to potential modified duty assignments and the physical requirements of jobs are valuable tools when faced with placement of an employee in a modified duty position, during the post-treatment meeting supervisors can be instrumental in helping the workers' compensation claims coordinator identify a modified duty assignment after an employee has been injured and released to return to work with a temporary physical restriction. Additionally, if supervisors are included in this post-treatment meeting, they are immediately informed of an employee's temporary physical restrictions.

Furthermore, for the purpose of promoting good employee/employer relations, suggestions from the injured employee regarding potential modified duty assignments should be taken into consideration during this meeting.

Once an applicable modified duty assignment has been identified, it should be documented and signed by all three parties involved in the post-treatment meeting. This quasi-contract should also identify the beginning and ending date of the modified duty assignment to preclude confusion and to enable the supervisor to adequately plan work within his or her area of responsibility.

7.5.4 Maintaining effective communication with the injured employee

Once an employee is assigned a modified duty position, the supervisor is critical to the ongoing success of that particular modified duty assignment. Because of the degree of interaction that supervisors have with employees, they should be charged with the responsibility of communicating the company's concern for the employee's well-being and with ensuring that the employee knows his or her responsibility to avoid violating the physician's restrictions. Furthermore, supervisors should be charged with the responsibility of observing employees to ensure that physical restrictions are not violated and should ensure that the workers' compensation claims coordinator is aware of the status, condition, and progress of each employee working in a modified duty position.

7.5.5   Maintaining effective communication with the insurance carrier

The workers' compensation claims coordinator is the company representative responsible for maintaining effective communication with entities outside the organization relative to workers' compensation claims. This is an exchange of information which includes both advising and soliciting information from outside entities to ensure adequate knowledge of pertinent issues. Because the management of workers' compensation claims is a combined effort between the employer and its workers' compensation insurance carrier, it is imperative that the workers' compensation claims coordinator maintain adequate communication with the claims representatives of its workers' compensation insurance carrier. This includes communicating all details surrounding each loss event, the scheduled dates for physician visits, and the results or findings of medical treatments. With respect to modified duty, the workers' compensation claims coordinator should immediately advise the workers' compensation claims representatives of the work status of each claimant. For the purposes of determining eligibility for indemnity benefits and to assist in the management of claims, the insurance claims representative must know whether the employee has returned to full-duty work, has been instructed by the treating physician (in writing) to remain off-work, or has been instructed by the treating physician (in writing) to return to work with temporary physical restrictions.

7.5.6   Revise modified duty assignments as applicable

Frequently the employee's treating physician alters the physical limitations of an injured employee as the employee progresses toward recovery. The physical limitations of an employee may also be modified if the treating physician believes more restrictive limitations than previously issued would more appropriately facilitate injury recovery. Since the physical limitations of an employee assigned to modified duty may change, it is necessary for the workers' compensation claims coordinator to be aware of them and to alter the modified duty assignment accordingly. For this reason, a meeting similar to the initial post-treatment meeting should be conducted with the injured employee, the workers' compensation claims coordinator, and the appropriate supervisors whenever an employee assigned to a modified duty task is evaluated by a physician. If changes are made to the original modified duty assignment, another quasi-contract should be completed to identify those changes.

The ultimate revision of an employee's modified duty assignment is returning the employee to full duty with no physical limitations. This is done only when the period of time for which the physician-imposed physical limitations has expired, or the employee's treating physician has provided documentation indicating that the employee can return to work with no physical limitations.

7.6   Summary

Modified duty is one of the most effective tools used to contain the cost of workers' compensation claims. Although resistance exists, it normally stems from an insufficient understanding of the benefits of modified duty. Furthermore, the benefits far overshadow reservations to implementing a modified duty program. Modified duty reduces future workers' compensation premiums, promotes good employee/employer relations, speeds the return of employees to full duty, and discourages abuse of the workers' compensation system.

Although it is possible to employ modified duty with little or no planning, the process is made much simpler if certain preparatory measures are taken. These measures include creating functional job descriptions, identifying potential modified duty assignments, developing a written modified duty program, communicating with medical providers, and conducting employee training.

As workers' compensation carriers are generally strong proponents of modified duty, most are willing to assist companies that they insure to develop and implement a modified duty program.

chapter eight

Accident investigations

Contents

8.1   Introduction

There is much truth in the saying, "Accidents don't just happen; they are caused." With that in mind, there should be a focus on determining the true cause of each accident so that corrective action can be taken to prevent future similar occurrences. It is generally true that history (including loss history) often repeats itself. For that reason, failure to adequately counter the causes of each accident may itself promote future accidents.

The primary tool used for determining the cause of an accident is an accident investigation. Implemented as a stand-alone loss control tool, accident investigations will likely offer only limited results. However, combined with other management initiatives, accident investigations are one of the most effective methods of reducing work-related injuries. Therefore, accident investigations are an integral element of an effective injury prevention program.

8.2   Benefits of conducting accident investigations

Although the primary purpose of an accident investigation is to aid in the prevention of future similar incidents, the investigation itself does not prevent accidents. Instead, it is the implementation of corrective action, identified through an accident investigation, that can prevent future accidents of the same kind. Accident investigations can be viewed as a structured means to a desired end, but in and of themselves they also have many benefits.

8.2.1   Identifies root causes of accidents

The process of conducting an accident investigation causes one or more people to view each accident with a figurative magnifying glass. The process leads to a conclusion that one or more causative factors contributed to the accident. These factors often include both readily evident causes and less evident or underlying ones. However, in the absence of a structured accident investigation program, the less-evident underlying causes of an accident are often overlooked because the traditional gathering of facts attributes many accidents to the conditions that are quick and easy to identify.

An example of this situation is an employee who is injured after slipping and falling in a puddle of oil. In the absence of an accident investigation program, the corrective action identified to prevent future similar occurrences may focus on cleanup of the spill. A structured accident investigation of the same accident, however, would likely focus on an ineffective machinery maintenance program that may result in leakage of fluid from machinery.

In this text the term *surface causes* is used to refer to the readily evident causative factors of an accident and the terminology *root causes* is used for the often less-evident underlying causes.

8.2.2   Creates awareness

In addition to providing a structured process for the identification of root causes, an accident investigation creates awareness of the potential for similar occurrences. As other management tools, such as a structured hazard analysis or a workplace safety inspection, may fail to identify some potential or existing hazards, the investigation of an accident brings the reality to light that a similar accident may occur in the future. This awareness is the foundation on which prevention efforts are built.

To illustrate this benefit of accident investigations, consider a truck driver who is injured after being struck by a forklift in the dock area of a customer's facility. A hazard assessment may have failed to identify this very real hazard. An accident investigation, however, raises an awareness of this hazard. Furthermore, documentation of the accident investigation provides a written record which will maintain awareness of this hazard beyond the limited tenure of the management representative who conducted the investigation.

8.2.3   Evidences trends

Even when management is aware of the potential for a particular type of accident, a detailed accident investigation program that categorizes its findings may identify trends that could otherwise go unnoticed. As an example, consider a company in which hand lacerations from utility knives is one of the primary causes of injuries. If that company does not conduct accident investigations, no significant trend may be noticed other than the fact that lacerations from utility knives are a primary source of injury. However, if that same company conducts routine accident investigations, it might niotice that 60% of these injuries occur immediately after a lunch break. By noting trends, the company may be better able to identify the root causes of these injuries. With the above example, one of the root causes may relate to fact that employees in that facility are sluggish and/or less attentive immediately following lunch breaks.

8.2.4   *Stimulates thoughts relative to prevention*

Accident investigations stimulate thought directed at the prevention of future similar occurrences. Indeed, it would be difficult to go through the steps of an accident investigation, including documenting the scene, obtaining witness statements, and identifying surface and root causes without following a logical progression toward recommendations for addressing the found causes. In short, merely conducting accident investigations stimulates the individual conducting the investigation to consider preventive measures. This benefit is multiplied when several people are used in the investigative process. The result is a brainstorming session which targets the prevention of injuries.

8.2.5   *Demonstrates management commitment and concern*

Although the benefits of accident investigations presented thus far would likely be considered direct benefits, the management commitment and concern exhibited by timely, thorough, and competent accident investigations are more of a collateral benefit. Although ancillary, the importance of employee perception of management should not be casually dismissed.

Any employee involved in a work-related injury serious enough to necessitate medical attention likely views the event as traumatic. An employer's efforts to investigate the cause of that injury naturally exhibits concern. Similarly, the employer's willingness to investigate accidents of less severity, including near-miss incidents, not only demonstrates concern but shows the employer's commitment to occupational safety at all levels.

8.2.6   *Identifies weaknesses in the safety program*

The goal of safety programs is to address all hazards that have the potential to cause employee injuries. However, the potential to have weak elements in individual safety programs and the potential to be missing programs that address relevant hazards are indeed real. The adage holds true that "a chain is only as strong as its weakest link."

As indicated earlier, conducting accident investigations causes management to identify the causes of each accident. By giving due consideration to the root cause of each accident, shortcomings in the company safety program frequently surface, and then the company safety program can be modified so that safety efforts are properly directed.

As an example of how accident investigations can expose weaknesses in a company's safety program, consider an accident in which a truck driver fell from the cab while exiting (a far too frequent occurrence). An accident investigation identified a root cause of the accident to be that drivers are not routinely trained to properly mount and dismount truck cabs. Although the company may require drivers to submit to random drug-testing, conduct pre-trip vehicle inspections, and attend driver safety meetings regularly, this type of training had been overlooked. The accident investigation, however, exposed the deficiency within the company's safety training program.

8.2.7 Justifies expenditures

To some degree, all companies work within a budget. Some allocate a predetermined amount of capital to each functional division of the company for both routine and nonroutine costs. Other, generally smaller entities, address nonroutine expenditures on a less formal, as-needed basis. Regardless of how companies addresses nonroutine disbursements, all work with a finite source of funds. For this reason, the ability to justify a particular expenditure may be the difference between whether or not the people "holding the purse strings" allocate the necessary funds.

By conducting accident investigations, the causes of each accident are identified and some form of corrective action is recommended. Adequately addressing the causes of an accident frequently involves an outlay of funds for such things as training materials, wages for employees during training periods, new equipment, or equipment maintenance. Justification for these expenditures may be based on a single injury accident or by a trend of accidents demonstrated through documentation and cumulative analysis of all accidents within a given period of time.

To illustrate how accident investigations can help justify expenditures, consider a trend of accidents in which employees have sustained back injuries from manually handling drywall when delivering it to job sites. As a means of avoiding similar injuries, the delivery supervisor proposes the purchase of a truck-mounted boom that would enable delivery personnel to place entire stacks of drywall in the openings of the structures to which they were being delivered, thereby eliminating almost all manual handling of drywall. Although past requests for this expensive device were once dismissed by the company owner as too costly, the trend of back injuries, which accident investigations attributed to the manual handling of drywall, provided justification for the expenditure.

8.2.8 Confirms or refutes compensability

In the absence of accident investigations, the process for submitting a workers' compensation claim consists of completing a form provided by the workers' compensation carrier and submitting it for payment of medical expenses and (if applicable) lost wages. This form provides the workers' compensation carrier only the minimal information necessary to process the claim. It provides very little information to help the claims representative determine if the injury is compensable.

Conversely, a thorough accident investigation provides detailed information, which may include witness statements, sketches, and photographs. The more information that is provided, the better able are the claims representatives to accept or deny a claim based on the applicable laws in that state.

For example, consider a nursing assistant who alleged a back injury from assisting with the transfer of a patient from a bed to a wheelchair. Without an accident investigation, this claim would likely be paid without question. However, if an accident investigation is conducted, and a witness indicates that the claimant did not appear to have been injured and did not say anything about being in pain, there may be some question concerning the legitimacy of the claim. There may not be sufficient cause to deny the claim, but it should cause both the employer and the workers' compensation carrier to more closely scrutinize the claim.

8.2.9 Provides evidence for the subrogation of a claim

Just as accident investigations provide information to the workers' compensation insurer which enables claims representatives to confirm or refute compensability, accident investigations also provide the workers' compensation insurer with information that helps determine if subrogation — shifting financial responsibility from the insurer to a third party — of the claim is a viable option. Relative to workers' compensation, subrogation is enabled when the actions or product of a third party cause a work-related injury. Examples of claims that can generally be subrogated are traffic accidents in which the operator of the other vehicle is at fault and injuries sustained as a result of a malfunctioning machine (assuming the machine had been properly maintained).

Many claims are not subrogated even when there is justification because in the absence of an accident investigation insufficient information is provided. To illustrate how an accident investigation can provide evidence for the subrogation of a claim, consider an injury in which an employee's fingers are amputated after being caught in a power press. If an accident investigation is not conducted, the injury description provided on the workers' compensation claim form may indicate "fingers caught in power press when removing part." Obviously the form provides the workers' compensation claims representative little or no information suggesting the viability of pursuing subrogation. Conversely, a thorough accident investigation with witness statements, sketches, photographs, maintenance records, etc. may provide needed information and enable the workers' compensation carrier to seek subrogation from the machine manufacturer.

It is important to note that subrogation of a claim can affect future premiums. As indicated previously, workers' compensation premiums are adjusted with respect to the frequency and severity of past claims. If the above injury costs the workers' compensation carrier a considerable amount in medical expenses, indemnity payments, and a permanent partial disability settlement, it may significantly raise the company's future workers' compensation premiums. However, if the claim is subrogated and a third party (such as the machine manufacturer) reimburses the workers' compensation insurance carrier, the severity of that claim would be reduced to $0, negating its impact on future premiums.

8.2.10 Reduces future workers' compensation premiums

Although subrogation of a claim can have a positive impact on future premiums, there is a more direct and more common way to reduce premiums through the use of accident investigations. As stressed throughout this chapter, the primary intent of accident investigations is to identify the causes of each accident so that measures can be implemented to prevent future similar occurrences. If an accident investigation program effectively aids in the prevention of injuries and workers' compensation claims, both the frequency and severity of claims are being addressed, which has a positive impact on the company's bottom line.

8.3 Focus of an accident investigation

As with any endeavor, it is important to begin with identifying the goal. Accident investigations are certainly no exception to this rule. The primary goal of every accident investigation should be to identify the causes of the accident so similar accidents can be prevented. Since both surface and root causes contribute to accidents, both are equally important to identify and address.

Several ancillary benefits of accident investigations can be viewed as secondary goals, including providing sufficient information to help the insurer determine compensability and viability of subrogation. However, humanitarian concerns outweigh the benefits of premium savings, and the primary goal of identifying and countering the causes of each accident is, and should remain, paramount. An accident investigation that fails to keep that goal in view is destined to fall short of its potential benefit.

The detriment of some accident investigation programs is that the people conducting the accident investigation lose sight of the goal and replace it with an effort to assign blame or find fault with individuals. Perhaps it is human nature to look for the easy answer and attribute an accident to the carelessness or inattention of an individual. However, assigning blame for an injury not only fails to identify and correct the causes but also creates an adversarial relationship between employees and management.

If the proper focus is maintained, an accident investigation program will be simultaneously reactive and proactive. It will be reactive whereby it is initiated only when there has been an accident or a near-miss incident. However, the proactive attribute is clearly evident when the implementation of corrective action eliminates one or more causes of future accidents.

8.4 Deciding which accidents to investigate

To some degree, all injuries and near misses should be investigated. Some companies limit accident investigations to only injury-producing accidents, and others limit them to injury-producing accidents that they deem serious. These companies obviously operate on the assumption that future occurrences of similar incidents will result in no injury or in injuries deemed less than serious and undeserving of the effort involved in conducting an accident investigation. These are faulty premises. The truth is that corrective action implemented in response to a near-miss incident may prevent a serious future accident. The FAA operates on this principle: it requires American-based airlines to report all near-miss events. By countering the causes of near-miss incidents, the U.S. airline industry has achieved remarkable success.

Although all accidents and near-miss incidents should be investigated, not all accidents and near-miss incidents deserve the same degree of investigative effort; detailed drawings, photographs, and the like may not be necessary for all accidents and near-misses. However, the degree to which investigations are conducted should not be based on the degree of injury that was sustained. Instead, this decision should be should be made on the potential severity of an injury. For example, if an employee lacerates his or her finger while using a utility knife, the injury may be treatable with only first-aid rather than professional medical treatment. However, it is conceivable that an employee performing the same task could have sustained a deep wrist laceration, severing a major artery. Therefore, because of this *potential severity*, the investigation following the finger laceration injury should be as thorough and diligent as if the deep laceration to the wrist had occurred.

8.5 Who should investigate

The task of performing accident investigations should not simply be assigned to the individual who willingly accepts the challenge. It is an important function with the potential to prevent employee injuries and to lower workers' compensation costs. Therefore, selection of the person or people charged with that responsibility should be given due consideration.

The first consideration is if responsibility for accident investigations should be given to one employee or a team of employees. Many companies use a single employee, such as a safety director or workers' compensation claims coordinator, but consideration should be given to using the team approach. A group of employees are more likely to think of more questions to ask during the fact-gathering phase, to consider a wider range of possible causes, and to contemplate more options for addressing the identified causes.

There are also specific considerations to be given to the individuals selected. In considering which employees will conduct accident investigations, attention should be given to the personal attributes of employees. Conducting accident investigations requires communicating with employees and gaining their cooperation, so anyone conducting investigations should be able to communicate with witnesses and injured employees effectively. Some people are uncomfortable interacting with others with whom a bond has not been established, while others have a coarse demeanor and may inadvertently cause friction when attempting to uncover details surrounding an accident. Individuals who presumably would have greater success conducting accident investigations are those who are described as empathetic and good listeners.

Other attributes desirable in those who conduct accident investigations are being goal- and detail-oriented, impartial, and objective. As discussed previously, an accident investigation program or investigator that loses sight of the primary focus and begins to address accidents by assigning blame is destined for failure. Conversely, an impartial and objective person driven to attain of a defined objective would likely be an ideal candidate.

In addition to personal attributes, the investigator should have detailed knowledge of the operation and tasks involved. If a single person or group of people within an organization possess sound knowledge of the entire operation, then one or more from that group could aptly be assigned the task of performing accident investigations. However, it is quite acceptable to identify individuals in each functional division of the company to perform accident investigations for accidents and near-misses that occur within their area of knowledge.

Last, the investigator should know how accident investigations are conducted. If a person such as a safety manager is already employed and has the requisite knowledge of how to conduct an accident investigation, he or she may be an ideal candidate, assuming that the aforementioned attributes are similarly possessed. If such a person does not exist, it is advisable that formal training relative to accident investigations be provided to the selected employees. This training can generally be obtained free through the loss control department of the workers' compensation insurance carrier.

8.6   Gathering information

8.6.1   Be prepared

The mere definition of an accident as an unplanned event suggests that accidents do not occur at expected times or locations. For this reason, it is imperative that accident investigators be prepared to conduct an accident investigation without prior warning.

To be properly prepared there should be a defined procedure for conducting accident investigations. This plan should address, among other things, which forms to complete, where to conduct the investigation, what authority the investigators have to interrupt the normal course of business to conduct an accident investigation, how and when to take photographs, make sketches, or take witness statements, and to whom and how the information should be forwarded.

In addition to having a predetermined plan of action, accident investigators should have the materials that are necessary in a single location from which they can be readily obtained. These materials will include such things as a camera, film, measuring tape, necessary forms, blank paper, pens, and caution tape.

8.6.2　Investigate with urgency

Ideally the accident investigation process will begin as soon after the accident as possible. Obviously, providing for the treatment of an injured party is paramount. However, of almost equal importance is securing the scene to prevent other injuries and to preserve the scene of the accident until it has been adequately documented. Securing an accident scene can be facilitated by the use of caution tape or by simply posting an individual at the scene to instruct others not to disturb the area until the accident investigator has finished documenting the accident scene.

The urgency with which accident investigations should be conducted is necessitated by several factors. First, witnesses to the accident are much easier to identify and easier to locate immediately following an accident. Second, the facts surrounding the accident are still fresh in the minds of the injured employee and the witnesses. Third, if accident investigations are conducted immediately following an accident, the witnesses do not have the opportunity to discuss among each other the events surrounding the accident. Given the opportunity to discuss the events of the accident before providing an independent statement to the person conducting the accident investigation lessens the chance that each witness will provide a true and unaltered description of what he or she witnessed. Finally, an accident investigation should be conducted with urgency because of the potential for the physical evidence to be altered. Environmental factors, such as lighting, noise, and temperature, may have affected the accident but may not be accurately recorded if an accident investigation is not conducted immediately. Furthermore, cleanup following an accident often destroys evidence that may be useful in determining the accident cause.

Although unrelated to the effectiveness of an accident investigation, conducting accident investigations immediately sends a clear message to employees that the well-being of employees and the prevention of future accidents are of utmost importance to the company.

8.6.3　Using accident investigation report forms

Because no form is thorough enough to ascertain all of the pertinent information for all workplace accidents and near-misses, conducting an accident investigation should not consist solely of completing an accident investigation report form. However, the use of accident investigation report forms is not without merit.

Perhaps the clearest benefit of accident investigation report forms is that they provide a standardized format for recording the information that is applicable to all work-related accidents and near-misses. It not only ensures that key questions are answered in every accident investigation but also requires less effort when reviewing the findings at a later time.

Another key benefit of standardized forms is that they enable the categorization of a predetermined series of factors. This categorization enables later analysis of cumulative data with relative ease to determine if there are any significant trends. Accidents can be categorized by the

- Nature of the injury (e.g., contusion, laceration, sprain, etc.)
- Injured part of the body (e.g., arm, finger, head, etc.).
- Manner in which the injury occurred (e.g., fall from elevation, being caught in machine, lifting, etc.)
- Employment longevity of the injured employee
- Department or location in which the injury occurred
- Day of the week the injury occurred.

In fact, the items that can be categorized on an accident investigation form are left only to the imagination of its creator. The only caveat is that the information categorized must be relevant.

Furthermore, since anyone interested in viewing the findings of an accident investigation may not need to know many of the details obtained during the investigation, the synoptic characteristic of a standardized report form is itself a benefit.

8.6.4   Documenting the accident scene

The first phase of an accident investigation is the documentation of the accident scene. The goal of documenting an accident scene should be to enable the re-creation of all pertinent characteristics of the scene at a later time from the written observations, sketches, photographs, and videotape.

8.6.4.1   Written observations

Written observations refers to the documented observations made by the person conducting the accident investigation on arrival at the accident scene. These written observations are of both physical and environmental conditions. Physical conditions to be documented would likely include identifying the tools and equipment at the scene of the accident, presence or absence of machine guarding devices, apparent presence or absence of personal protective equipment, names of the people present at the accident scene, and any other physical conditions that may have contributed to the accident, such as a wet, slippery, or uneven floor surface. Environmental conditions to be documented would include the time of day, temperature, lighting conditions, any excessive noise levels, and any unusual odors.

8.6.4.2   Sketches

Sketches of accident scenes provide details that are too difficult to describe concisely with written observations. Furthermore, sketches often reinforce written observation of physical conditions and are ideal for enabling the reconstruction of accident scenes, as they frequently include measurements.

It is not necessary to provide detailed illustrations in accident scene sketches, as photographs and/or videotapes provide the needed detail. Hence, the investigator need only provide a rough outline of each relevant object and a label that identifies each object on the sketch. In addition to the physical objects contained in an accident scene sketch, the location of each witness at the time of the accident and the location from which each photograph was taken should be included.

Graph paper is ideal for accident scene sketches, as each square can be estimated as a standard unit of measure (e.g., one square is roughly equivalent to one foot). However, when locating an object on a sketch using measurements, the process of triangulation should be used. This process simply involves identifying the location of each movable object that is pertinent to the accident scene by measuring its location from two fixed objects. On a sketch, triangulation should be represented by arrows which denote the measured distance between two points. If the position of an object is critical, as opposed to its mere location, triangulation should be performed for two opposite points of the object.

8.6.4.3 Photographs

The saying "a picture is worth a thousand words" certainly holds true for accident investigations. One press of the shutter button on a camera can save an immeasurable amount of time and effort when compared with trying to provide written documentation of every relevant aspect of an accident scene. However, accident-scene photography is more involved than taking snapshots on a family vacation.

At the onset, the investigator should predetermine the photographs that are necessary. If the scene is complex, the investigator would be well-served to first list the photographs to be taken. These photographs should include a general overview of the accident scene, as well as midrange and close-up photographs. Additionally, each relevant subject of accident scene photographs should be taken from several different angles. Furthermore, it is often beneficial to include an object, such as a rule, in each photograph that depicts the scale of the objects contained therein.

The accident scene should be photographed prior to any alterations. However, it is an acceptable practice to make necessary alterations to the scene so that subsequent photographs show otherwise obscured features, insofar as the initial photographs depict the scene prior to moving objects or otherwise disturbing the scene.

Proper records should be maintained relative to each photograph. These include the frame number of the photograph, the location from which the photograph was taken, the identification of the photograph subject, and an indication of any modifications that were made to the accident scene for the purpose of providing more detailed photography.

8.6.4.4 Videotape

Using a video camera to document an accident scene adds several additional benefits to the use of still photography. First, videotaping an accident scene provides a virtually infinite number of photographic images without taking thousands of still photographs. Also, if a photograph is desired, technology is readily available (and relatively inexpensive) to make a photograph from virtually any frame of a videotape. Second, videotaping an accident scene provides the added benefit of having an audible component. Therefore, if videotape documentation is taken at the accident scene immediately following an accident, the investigator can not only capture visual documentation of the scene but can also record comments and conversations made by people at the scene. With the assumption that the videotaped evidence is obtained immediately following an accident, a third benefit is provided by panning 360 degrees around the scene, whereby potential witnesses can be identified through later viewing of the videotape. An additional benefit to videotaped documentation of an accident scene is realized if the accident scene must be altered to treat the injured worker or to prevent other employees from becoming injured. In such situations, videotape documentation accurately records how the accident scene has been altered.

8.6.5 Acquiring witness statements

After the accident scene has been adequately documented, it is time to acquire statements from the witnesses. In this context, the term *witness* applies not only to bystanders who may have witnessed the incident but also to the injured employee. In fact, frequently the only witness to an accident or a near-miss incident is the person who was injured or nearly injured. Because people are knowingly or unknowingly influenced by the statements of others, attempts should be made to keep witnesses separate until all have rendered statements, so that they do not have the opportunity to discuss the incident among themselves.

This phase of the investigative process is when answers to the questions of who, what, when, where, how, and why should be ascertained. These terms are starting points from which many relevant questions can be formed. For example, consider the numerous questions that can be posed by asking "what": What time of day did the accident occur? What did you observe? What did you hear? What were you doing immediately prior to the accident? What were you doing at the time of the accident? What safeguards were available? What safeguards were used? What training have you received relative to the task that you were performing? What preventive maintenance is performed on the machine? What knowledge do you have of any other similar injuries or near miss events? These are all open-ended questions and invite the witness to speak openly. Closed-ended questions, which require only an affirmative or negative reply, such as "Were you wearing safety glasses?" limit the interviewee's response. For this reason, closed-ended questions should be avoided when possible. Instead, witnesses should be encouraged to speak and should not be interrupted.

In addition to avoiding closed-ended questions and allowing the witness to speak uninterruptedly, the questions posed to the witness should be in such a manner as to solicit a description of the events in chronological order. This will make the interview proceed more smoothly and will make notes more easily understood at a later time. A simple approach to soliciting a description of events in chronological order is to first ask the witness what he or she was doing immediately preceding the incident and then continually ask, "What happened next?"

After the interview, the witness should render a written statement. The process of questioning the witness prior to asking for a written statement will likely have the effect of increasing the degree of detail within the written statement. As it may be used as a legal document, the written statement from each interviewee should be signed and dated by the person rendering the statement. For serious injuries, it is likely that the workers' compensation claims representative will request a copy of any written statements that were obtained.

While still at the scene of the accident, the investigator should get both verbal and written statements from the witnesses, including the injured employee. Caution should be taken in this phase of the investigation not to be coarse or appear accusatory. The investigator should remain objective throughout the entire process and should not arrive at conclusions hastily. As mentioned previously, the investigation process must have the primary and ultimate focus of determining the accident cause so that future similar occurrences can be prevented. This focus should be sincerely communicated with each witness, particularly the injured employee. The cooperation that is received will likely be proportional to the degree to which the interviewee feels that he or she is valuable in the effort to create a safer work environment.

Even if an accident investigation is not conducted immediately following an accident, all reasonable efforts should be made to acquire witness statements at the actual scene of the accident. This lends itself to a clearer description of how the accident actually occurred, as the witnesses are better able to convey what they saw and their location when they saw it, since they are able to point to a specific location or object. Witnesses, and particularly the injured employee, should also be asked to demonstrate how the accident occurred, if it can be demonstrated safely. This method provides an explanation of the event that words far too frequently are not able to convey adequately. In some situations, the attempt of an individual to demonstrate how the accident occurred may cause the investigator to realize that the accident could have not occurred in the manner in which the injured employee described. In such situations, the claims representative from the workers' compensation insurance carrier should be notified, as it raises a red flag relative to workers' compensation fraud.

8.6.6   Reviewing relevant documentation

Existing documentation can be of immeasurable benefit to helping accident investigators gain insight. Relevant documentation includes that which relates to existing safety efforts, established procedures, past accidents, product specifications, equipment maintenance, and specific employees.

Documents which reflect a company's existing safety efforts include written safety programs, safety rules, safety incentive programs, minutes from safety committees, documented safety inspections, and correspondence from outside entities such as OSHA compliance inspectors and workers' compensation loss control consultants. By reviewing these documents, the accident investigator is provided with a snapshot of the company's efforts. Additionally, by reviewing safety programs the accident investigator identifies the existing policies and procedures for preventing accidents and the established content and schedule for safety-related training. A review of these documents can identify weaknesses of the programs, which may help the investigator determine the root cause of the accident. Furthermore, documented findings from safety inspections, to include both internal inspections and audits conducted by outside entities, may reveal physical hazards that have not been mitigated or recommended managerial efforts that have never been implemented. Minutes from past safety committee meetings may include discussion which pertains directly to the prevention of specific accidents and the measures which have been enacted thus far in response to those discussions.

Virtually every company has established procedures for performing the tasks of each job. If these procedures are written, they can be reviewed as an element of the accident investigation process. Additionally, if training records are maintained relative to the standard operating procedures, then the content of training provided to employees, the date of the most recent training, and the identification of employees who have received specific training can be examined. From this information, the accident investigator may be able to identify oversights in the standard operating procedures, inadequate or ineffective training, or even employees who have seemingly slipped through the cracks with respect to receiving necessary training.

Documentation which provides information about past accidents may also be a benefit in the accident investigation process. These records may suggest accident trends, provide details regarding similar past accidents, and make the tasks of identifying the root cause of an accident much easier. This documentation includes past accident reports, past accident investigations, police reports, workers' compensation loss runs, and OSHA 200 logs (Summary of Occupational Injuries and Illnesses).

For accidents that involve products, equipment, or machinery, the accident investigator may want to review such documentation as the operator's manual, maintenance agreements, internal maintenance records, and product sheets such as material safety data sheets (MSDS). With respect to equipment and machinery, a review of this information will likely permit the accident investigator to determine if there are any inconsistencies between the manufacturer's intended operating procedures and the manner in which the equipment or machinery is customarily operated by employees. Additionally, a review of information will enable the accident investigator to determine if there have been any modifications to the equipment which could have an impact on its safe operation. Furthermore, a review of both internal and external maintenance records will indicate if the preventive maintenance has been provided at its intended frequency. Information sheets, such as MSDS, help the accident investigator to determine the specific hazardous properties of a product, the recommended personal protective equipment, and other pertinent information.

Yet another category of documentation that may be beneficial to review during an accident investigation is personnel data, such as past disciplinary reports, work schedules, and performance appraisals. By reviewing this information, the accident investigator may be able to determine if the potential causes for the accident being investigated included such things as exhaustion from working too many hours, an adversarial relationship with a supervisor that distracted the employee from performing his work safely, or insufficient motivation for an employee performing a specific task in a prescribed manner.

8.7   Identifying surface causes and root causes

Despite well-intentioned efforts, many accident investigations fail to provide the degree of benefit that they are professed to deliver. This is generally not because of improper motive or a lack of effort in the previously described phases of the accident investigation process. Instead, the primary flaw in many less-than-ideal accident investigations is the failure to correctly identify both the surface causes and the root causes of the accident being investigated.

Perhaps human nature is the limiting force that leads individuals to seek the easiest solution to any question. However, when the question is, "What caused the accident?," the person who seeks the quick and easy answer will likely fail to identify the root cause of the accident. To understand why this practice is substandard, it is necessary to thoroughly understand the terms *surface causes* and *root causes*. Figure 8.1 depicts their relationship.

8.7.1   Surface causes

Surface causes are the factors that have a direct impact on accident causation. They are often readily apparent and are often easy to correct once identified.

To illustrate surface causes, consider an employee who ran across the factory floor to answer the telephone and subsequently tripped and fell on a metal strapping band. In this illustration, the most evident surface causes of this accident are the metal strapping band left on the factory floor and the employee running. Although there may not always be more than one surface cause to every accident, as illustrated in this example more than one surface cause for a single accident is quite possible. Both of these surface causes were able to be identified with relative ease and are similarly easy to correct.

It is commonly said that work-related injuries are caused by either unsafe conditions or unsafe acts; these are surface causes. Hence, surface causes can be considered to be of two types — those relating to unsafe conditions, and those relating to unsafe acts.

8.7.1.1   Unsafe acts

Unsafe acts are the behaviors that directly contribute to the cause of the accident. With respect to work-related injuries, unsafe acts are frequently

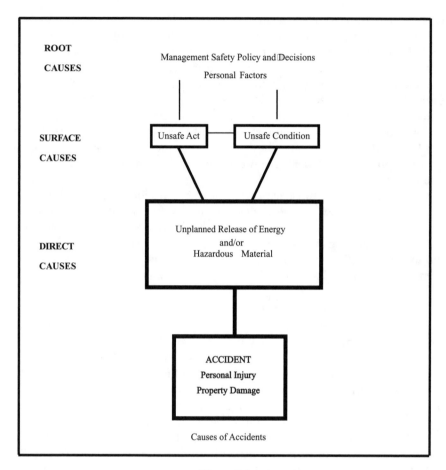

Figure 8.1

called unsafe work practices. They may involve the injured employee's action or failure to act, or they may result from the action or omission of another individual. Unsafe acts include such things as failure to look in the direction of travel when operating a forklift, removing a guard from a machine, using improper lifting techniques, failing to obtain assistance from another employee when warranted, and failing to use available material-handling aids.

Unsafe acts are frequently deviations from accepted safe work practices. However, it is important to note that an unsafe act is not always a deviation from the norm. Consider a work environment in which the company rules indicate that employees must wear safety glasses when using a pneumatic nail gun. However, in that workplace it is common practice to wear no eye protection while nailing. It is also common practice for supervisors and management to fail to correct employees without safety glasses using a nail gun without safety glasses. In such circumstances, the unsafe work practice is not a deviation from the norm, but that fact does not make failure to wear safety glasses any less an unsafe act or a surface cause of an accident.

Although the above illustration suggests that the employees who failed to wear safety glasses did so as a result of a conscious choice, unsafe acts are not always conscious decisions made by the injured employee or other individuals. It is indeed possible for an employee to commit an unsafe act as a result of ignorance. Whereas safety has been described as the application of common sense, there is no common sense without common knowledge.

In addition to conscious choices and ignorance, unsafe acts may be the result of human error. In this context, human error refers to misjudgments made by an individual. An employee who misjudges the weight of an object that he is lifting does not necessarily make a conscious choice to lift something that is too heavy. Nevertheless, the act is still categorized as unsafe.

Last, unsafe acts may be the result of the limitations of the human body, such as performing a task that involves a great deal of repetitive wrist motion. Performing that task is an unsafe act, as it may contribute to a cumulative trauma disorder.

8.7.1.2   Unsafe conditions

Unsafe conditions are the attributes of the workplace that directly contribute to the cause of an accident. They are also called hazardous conditions, as the existence of an unsafe condition creates a hazard. They may be either physical characteristics or environmental conditions.

Physical characteristics that may constitute unsafe conditions may include such things as a spill that creates a slip hazard or a poorly designed workstation that requires the employee to move in an awkward manner. However, physical characteristics that constitute unsafe conditions can include not only things that are present but also things that are absent. For example, the absence of an effective guarding mechanism on a power press constitutes an unsafe condition.

Environmental characteristics constituting unsafe conditions may include such things as excessive temperatures, excessive noise, poor lighting, and inadequate ventilation. Although these characteristics are not tangible features of a workplace, they nevertheless can cause unsafe conditions.

8.7.2   Root causes

Root causes are the conditions that allow the surface causes to exist and almost always have their foundation in the administrative and managerial systems of a company. As indicated above, the surface causes of accidents are generally categorized into either unsafe acts or unsafe conditions. By understanding why an employee acted in an unsafe manner or why an unsafe condition existed, the accident investigator is able to identify the root causes of accidents.

The most common and appropriate means of illustrating the relationship between surface and root causes of accidents is to consider every accident as a weed. The leaves of the weed are analogous to the existing unsafe conditions and work practices within a company. The roots of the weed represent the deficiencies within the company's administrative and managerial functions. Since they are underground, they are far less apparent than the surface causes.

As with an actual weed, the entire top of the weed can be removed and the weed will disappear from sight, only to return and flourish until it is ripped off at the base again. The same is true with the weed that represents an accident in the workplace. If an accident investigation identifies one or more surface causes that are then mitigated, it is likely that there will not be any immediate recurrence of the accident. However, accidents will surely recur if the underlying root causes are not countered. Conversely, just as removing the roots of a weed will cause the entire plant to wither and die, removing deficiencies within the administrative and managerial functions will cause unsafe acts and unsafe conditions to cease, including those which have not even been identified.

Many individuals charged with conducting accident investigations, but lacking a sound knowledge of accident prevention fail to inquire any further once a surface cause has been identified. By failing to identify and address the root causes of each accident, future accidents continue to occur. However, a lack of knowledge concerning the anatomy of an accident is not the only reason that many accident investigations are considered complete before the root cause has been identified. Identifying the root causes of accidents requires some uncomfortable introspection. Although no person enjoys finding deficiencies in their own actions, if managers or supervisors are charged with the responsibility of conducting accident investigations, they must accept the reality that no management system is flawless and that no flawed system can improve until the deficiencies are identified. Figure 8.2 identifies some of the deficiencies that can reside within administrative or managerial functions.

In response to identified surface causes that are categorized as unsafe acts, an investigator might ascertain if the employee knew the rules and procedures. If the employee was unaware of specific rules and procedures, why was that condition so? Was there training conducted? Was the training appropriate? Why was the training apparently ineffective? If the employee indicated that he or she was aware of the specific rules and procedures, the investigator should not stop seeking further explanation but should simply ask a different series of questions. Was there anything that inhibited the proper performance of the task? Was there motivation for performing the task in a manner contrary to the prescribed manner? Was there adequate supervision?

In response to identified surface causes categorized as hazardous conditions, an investigator might ascertain if the condition had been recognized previously. If the condition was not recognized previously, why not? Are there effective safety inspections being conducted? Are employees being provided with the means and incentive to report unsafe conditions? Is there

Deficiencies within Administrative Functions

- Absence of routine means of recognizing unsafe conditions
- Absence of routine means of recognizing unsafe work practices
- Failure to provide employee safety training
- Inadequacy of training (failure to address specific hazards)
- Failure to provide safety incentives and motivation
- Failure to hold supervisors and employees individually and collectively responsible
- Inadequate preventive maintenance emphasis
- Lack of due diligence in purchasing

Deficiencies within Managerial Functions

- Failure to correct unsafe work practices
- Failure to recognize unsafe work conditions
- Failure to motivate employees
- Lack of supervisor involvement in safety program
- Failure to adequately convey expectations of employees
- Failure to instruct employees in safe work practices
- Failure to instruct employees regarding hazards
- Failure to forward safety-related suggestions to management for approval
- Failure to provide adequate supervision

Figure 8.2

an effective preventive maintenance program? Is the work area or equipment poorly designed? Conversely, if the condition was recognized previously, what was done to address the hazard? What caused the hazard to recur? What root causes were previously identified as having contributed to the existence of the hazard?

8.8   Mitigating the causes

An accident investigation is not complete once the facts are gathered and the surface and root causes of the accident have been identified. An accident investigation also includes recommending and implementing measures to counter the causes that have been identified. Because surface causes and root causes have different characteristics, the mitigation of these hazards must be approached differently.

8.8.1   Addressing surface causes

There is a structured manner by which surface causes should be addressed. This process involves applying the strategies of engineering controls, administrative controls, and personal protective equipment in their prescribed order.

8.8.1.1 Engineering controls

Where feasible, engineering controls are the preferred means of controlling the unsafe acts and unsafe conditions uncovered by accident investigations. Engineering controls are physical changes to work environments that control exposure to hazards. They act on the source of the hazard and control employee exposure to it without relying on the employee to take self-protective action. Examples of engineering controls include changing, modifying, or redesigning workstations, tools, equipment, materials, and processes.

8.8.1.2 Administrative controls

If engineering controls are not feasible or do not adequately mitigate the hazards identified by an accident investigation, administrative controls should be implemented. These administrative controls need not be a stand-alone means of controlling the hazards but may be implemented in conjunction with engineering controls. Administrative controls are procedures and methods that significantly reduce exposure to hazards by altering the way in which work is performed. Examples of administrative controls include adjusting the pace of the work, redesigning how work is performed, instituting rest breaks, or implementing a schedule of job rotation.

8.8.1.3 Personal protective equipment

Because personal protective equipment does not eliminate the hazard but instead simply provides a barrier between the worker and the hazard, it is the least-preferred means of addressing unsafe acts and hazardous conditions. Furthermore, the use of personal protective equipment relies on employees using the equipment consistently. For these reasons, if personal protective equipment is used as the sole means of protecting employees from a hazard, both engineering controls and administrative controls should have previously been dismissed as ineffectual. However, personal protective equipment may be used in conjunction with engineering and/or administrative controls to provide an increased level of protection.

8.8.2 Addressing root causes

Although all accident investigations should include the identification and implementation of measures to alleviate the identified surface causes, it is even more important to identify and implement control strategies that address the deficiencies in the administrative and management systems that nurture the surface causes. Surface causes exist only as a consequence of deficiencies within the administrative and management systems. Therefore, by effectively addressing the root causes of each accident, a company is simultaneously addressing both evidenced and latent surface causes.

Although there is no prescribed hierarchy of control strategies to address deficiencies within administrative and managerial systems, the task of addressing these root causes is not without challenges. Instead, because implementing measures to address root causes requires a company's upper management to admit that the existing system may have flaws, it will likely be a more difficult undertaking than it appears. For this reason, gaining approval for making changes to the administrative and management systems necessitates preparedness. The accident investigation must clearly define the correlation between the surface and root causes. If possible, it should also identify other unsafe acts or conditions that could reasonably result from a failure to make modifications to the existing administrative or management systems. For control strategies representing substantial changes, it may be necessary to provide a cost-benefit analysis.

If the person conducting the accident investigations is not a part of upper management, it may be beneficial for the accident investigator to gain the assistance of a person outside the company in conveying the need for changes in policy, training, purchasing, accountability, etc. An ideal source for this type of assistance is the loss control staff of the company's workers' compensation insurance carrier, as a loss control consultant will likely be perceived as an individual with a single motive, the reduction of accidents and corresponding reduction in workers' compensation claims. A loss control consultant employed by the company's workers' compensation carrier may also be able to describe how similar changes within other companies have produced desirable benefits.

8.9   Evaluating control strategies

Once specific control strategies have been implemented to address both the surface and root causes of each accident, only one phase of the accident investigation process remains. This phase is the evaluation of the strategies that have been implemented. It would be wonderful if we could look into a crystal ball and forecast if the control strategies will be effective. However, this is not reality. Accordingly, it is important to examine the implemented control strategies to determine if they are providing the desired result. An evaluation of the control strategies also shows if the application of those strategies is consistent.

8.10   Summary

As is the theme of this text, an accident investigation is an effective tool to counter the tremendous cost of work-related injuries and workers' compensation insurance. By interceding in the cycle of accidents, similar accidents will be averted and the incurred losses will continually be reduced. However, conducting accident investigations that produce results requires that they be conducted thoroughly and consistently. The investigation must properly identify both the surface and root causes of each accident and near-miss incident. The identified causes must be countered with effective corrective action, and the effectiveness of the corrective actions must be assessed to ensure that the intended goal is being attained.

As with each of the management tools presented in this text, accident investigation is not a stand-alone cure-all for work-related injuries and high workers' compensation costs. Just as a building cannot be constructed using a single tool, neither can an effective workers' compensation cost control program be fashioned with a single management tool.

chapter nine

Substance abuse programs

Contents

9.1   Introduction

Today in the U.S., 71% of all drug users over the age of 18 are employed either full- or part-time; that's more than 10 million workers. The chances that your company employs a substance abuser, regardless of the size of your business or the number of employees you have, is greater today than it has been in the past several years. Why? Because substance abuse in America is on the rise, and it hasn't left the workplace out of its path of destruction. The current high employee turnover rate greatly increases these chances, as well.

What can employers do to curb the growth of substance abuse, including drugs and alcohol? This is a critical question to ask because if you haven't done anything yet, chances are you have a bigger problem than you realize. Studies reveal that substance abusers as employees have a tremendous effect on the workplace and cost their employers. For example, employees who abuse drugs are less likely to show up to work on time or put in a productive day's work. They are more likely to be absent from work without a legitimate reason, use their healthcare benefits, and file workers' compensation claims.

According to the National Council on Compensation Insurance, as many as 50% of all workers' compensation claims are related to the abuse of alcohol or drugs in the workplace. Drug users, as a group, use medical benefits at a rate eight times higher than nonusers. Substance-abusing employees also are absent from work more often than their nonusers, often resulting in increased workloads for coworkers and decreased employee morale.

What's the bottom line? Substance abusers in the workplace significantly contribute to increased healthcare costs, disability insurance costs, absenteeism rates, employee theft, and accidents, as well as decreased productivity, product quality, and employee morale. Though there is no exact figure, reports estimate that substance abuse is costing American businesses billions of dollars each year.

Unfortunately, once employers begin to see the dollars dropping from their bottom line, it is often late in the development of a serious substance abuse problem. Can this be avoided? Can employers look for signs of possible substance abuse? Are there hidden signs that might warn employers that a problem is developing? The answer to all these questions is yes. In most cases, many of the long-term problems associated with workplace substance abuse can be avoided if employers are constantly on the lookout for the signs of substance abuse and take appropriate action. However, it is also important to recognize that an employee's declining job performance may be caused by factors unrelated to substance abuse. Supervisors should be trained to measure each worker's job performance and to refer employees with performance problems to the appropriate qualified professional to determine the nature of the problem.

9.1.1   Employee performance

There are several ways to measure workers' performance that can also help employers spot potential substance abuse problems early on. Is a certain employee's quality of work inconsistent? Is the employee's work pace slow, slower than usual, or sporadic? Does the employee have difficulty concentrating on work? Are there signs of fatigue? Other telling performance signs include mistakes, errors in judgment, and a sudden inability to fulfill complex assignments or meet deadlines. Increased absenteeism or tardiness, both of which have a direct impact on the performance of the troubled employee and the coworkers who have to carry the extra workload, also could indicate that a substance abuse problem exists. Other performance-related signs of substance abuse may include

- Excessive sick leave
- Frequent early departures
- Patterns of absenteeism (Mondays, Fridays, the day before or after holidays, and the period following paydays)
- Extended coffee breaks
- Excessive time on the phone

9.1.2 Behavior and appearance

Workers who display sudden changes in behavior on the job may be trying to hide a substance abuse problem. For example, irritability, moodiness, arguing with coworkers, or insubordination toward supervisors is not uncommon among substance abusers. For substance-abusing employees, personal appearance may lose its usual importance. Troubled workers will often come to work sloppy, unkempt, unshaven, or dressed inappropriately (long-sleeved shirts in the summer, sunglasses indoors, etc.).

Also, employers may begin receiving complaints from customers, clients, and coworkers regarding the attitudes and work quality of substance-abusing employees. Family members may even contact the employer seeking intervention. Several additional behavior-related signs of substance abuse are listed in Figure 9.1. It's important to note, however, that the presence of one or some of these signs does not unequivocally indicate that substance abuse is occurring.

Substance-abusing employees are not safe employees. Depending on the type of work employees do, substance abuse problems can begin affecting employee safety records. Substance-abusing employees will be involved in more accidents than other workers, even though they are often not the ones who are injured. They also tend to display carelessness in the operation and maintenance of potentially hazardous materials or dangerous equipment. Other safety-related signs of substance abuse may include

- Risky behavior
- Increased involvement in off-the-job accidents
- Damaging equipment or property

9.2 Benefits of substance abuse programs

There are many demonstrable benefits of workplace substance abuse programs, including a decrease absenteeism, disciplinary problems, and

• Sleepiness	• Stealing from the company and coworkers
• Slurred speech	
	• Sudden change in choice of friends
• Unsteady movements and shaky hands	
	• Poor personal hygiene
• Cold, sweaty palms	
	• Violent behavior
• Dilated pupils	
	• Impatience
• Red, bloodshot eyes	
	• Depression
• Unusual weight loss or gain	
	• Suspicious attitude towards others
• Smell of alcohol on breath	
	• Emotional behavior
• Deteriorating family relationships	
	• Excessive talkativeness
• Borrowing money from coworkers	
	• Withdrawal from coworkers

Figure 9.1

employee turnover, an improved work quality and morale, and a better general health status of the workforce as a whole. Other potential benefits of such a program include increased productivity and improved community image. Finally, workplace substance abuse programs provide benefits that both directly and indirectly impact the costs relating to workers' compensation insurance.

9.2.1   Premium discounts

Perhaps the most direct and immediate impact that workplace substance abuse programs have upon workers' compensation costs is a direct discount offered by insurance carriers to employers who have a workplace drug-testing program. Although all workers' compensation insurance carriers do not offer this premium credit, it is notable that many carriers do. The carriers who offer this credit often have guidelines with which workplace drug-testing programs must comply for the discount to be applicable. When offered by an insurance carrier, a credit applied for having a drug-testing program takes the form of a percentage discount applied to the employer's modified premium. As such, an employer can save money simply by implementing a drug-testing program, even if the program fails to yield any other evident benefits. Depending upon the amount of the credit offered by an insurance carrier, the discount for simply having a workplace drug-testing program may completely offset the cost of administering the program.

9.2.2 Compensability determinations

Whereas the premium discount offered by many insurance carriers is likely the most immediate and most direct manner by which workers' compensation costs are impacted by workplace drug-testing programs, it is not the sole benefit. In addition, drug-testing programs can provide workers' compensation insurers with the ammunition that is necessary for denying injury claims that are not compensable. Workers' compensation benefits are denied when the claimant is injured as a direct result of being under the influence of alcohol or illicit drugs, but the insurance carrier needs some form of quantifiable evidence to prove the condition of the claimant at the time of his or her injury. Drug-testing programs, which include the testing of employees immediately following any work-related injury, provide that evidence. Although post-accident drug-testing is the only manner by which noncompensable claims can be denied on the basis that the intoxicated state of the claimant led to the injury, it is important to note that most state legislation requires the insurance carrier to show a causal relationship between the intoxicated state of the claimant and the incident that caused injury. For example, a claim would most likely be denied for an intoxicated outside salesperson who was involved in a traffic accident, since there is a logical casual relationship between the salesperson's intoxicated state and the resulting injury-producing incident. However, a claim for an intoxicated office administrator who was injured when a vehicle crashed through the wall of his office would likely not be denied, since the intoxicated state of the office administrator did not cause the vehicle to crash through the wall of his office.

9.2.3 Fraud prevention

It is noteworthy to draw the correlation between illicit drug use and workers' compensation fraud. By their very definitions, both are violations of criminal law. Although perhaps supported only by anecdotal evidence, a person who lacks the moral character to refrain from illicit drug use may also lack the moral character to refrain from committing workers' compensation fraud. The same logic may be used to rationalize both, as some may have the faulty view of both illicit drug use and workers' compensation fraud as victimless crimes. If through a drug-testing program, which includes preemployment testing of applicants, an employer is able to identify and deny employment to individuals who use illicit drugs, potential future incidences of fraud may be simultaneously averted.

9.2.4 Accident prevention

Perhaps the primary benefit of substance abuse programs is their potential to influence the number of work-related accidents. Alcohol and other drugs in the workplace can create problems that may jeopardize the safety of employees. Although drug-testing programs are merely one tool used to reduce the number of employee injuries within a given workplace, the wide acceptance of drug-testing programs is a testament to the confidence in their value. In fact, through the Federal Drug-Free Workplace Act and Department of Transportation regulations for drug-testing individuals in safety-sensitive positions, drug testing has likely become the most widely implemented management tool for the prevention of work-related accidents.

When drug-testing programs are successful at preventing work-related injuries, the frequency of workers' compensation claims is reduced. Because workers' compensation insurance premiums are based partly on the frequency of claims incurred over several years, the ability of a drug-testing program to reduce workers' compensation claims translates into lower workers' compensation premiums.

9.3   Applicability of drug testing

Smaller employers often feel that their businesses are not susceptible to the problems of alcohol and drug abuse among employees. The reasons given are often centered on the employers' belief that they know their employees and that none abuse alcohol or drugs. However, perceptions are not always correct. Furthermore, in a company with only ten employees, just one with an alcohol or other drug problem can be devastating to company efforts to prevent accidents and reduce workers' compensation costs. According   to the Center for Substance Abuse Prevention (CSAP), in the booklet entitled *Creating a Drug-Free Workplace*, drug users, at a minimum, consume almost twice the medical benefits as nonusers, are absent 1.5 times more often, and account for more than twice as many workers' compensation claims.

Drug and alcohol testing do not constitute a substance abuse program. Many companies, however, believe that when combined with the other components of a comprehensive substance abuse program, testing can be an effective deterrent to substance abuse and an important tool to help employers identify workers who need help.

Though setting up a testing program is not a simple process, every year more and more companies of all sizes are doing so. Some establish programs because state or federal laws or regulations require them to. Others test to take advantage of incentive programs made available through the state or an insurance provider. Still others do so because it is the right business decision for the company.

Before you implement a drug- or alcohol-testing program, consider the following questions:

- Whom will you test? (job applicants, all employees, selected employees, employees only at certain job sites)

- When will you test? (after all accidents or only after some, when you suspect that an employee is using drugs, as part of periodic physical examinations, randomly)
- For what substances will you test? (only the five drugs required by many federal government agencies — marijuana, opiates, amphetamines, cocaine, and PCP; only marijuana and cocaine because they are the most commonly abused illegal substances; alcohol because it is the most-abused substance in American workplaces; other legal substances that are commonly abused (such as prescription drugs) and that can affect job performance)
- What are the consequences for employees and applicants if they test positive?
- Who will administer your testing program?

Drug testing has been gaining in popularity in the private sector for the past decade. During that time, many safeguards and confidentiality measures have been developed to ensure the quality and accuracy of drug testing. In addition, laws and regulations govern how programs must be established and conducted. Before implementing a testing program, you should contact an individual or organization with expertise in drug- and alcohol-testing issues to help you establish your program.

9.4   Creating a clear policy

The need for policies that define the company's position and its expectations of employees is widely accepted. Most companies, regardless of size, have policies regarding attendance, conduct, and even smoking in the workplace. Furthermore, this test promotes the imlementation of policies that address company efforts to control the costs relating to workers' compensation insurance, such as a those relating to modified duty and accident investigations. Therefore, the first and most important step to address the problem of alcohol and other drugs in the workplace is the creation of an alcohol and drug policy that clarifies the company's position regarding alcohol and drug use in the workplace.

Before you actually start writing your substance abuse policy, there are a number of important steps that you may want to take. A needs assessment survey may help you to better understand the company's current situation and determine exactly what you want the program to accomplish. Enlisting the assistance and input of your employees will not only help you to develop the best policy possible but also will help secure your employees' support. Your workers should be your allies in this effort.

A well-designed policy should explain why such a policy is needed and its importance. Simply describing the goal of the company's drug-testing program can facilitate this. The policy should also clearly convey that alcohol and other drug use on the job violates company policy and will not be tolerated. Furthermore, the policy should state the specific consequences for violating the policy and should define how, when, and under what circumstances employees will be tested. Lastly, the policy should state specific ways for employees and family members to obtain help for alcohol or drug abuse and their related problems. A sample policy is provided in Figure 9.2.

9.5 Involvement of the supervisor

The level of support your supervisors give to the company's substance abuse program, combined with the fairness of your program and the firmness of your commitment, will greatly influence its potential for success. Many of the problems encountered when implementing a program can be avoided if you have the full support and participation of your supervisors and managers.

Your supervisors are responsible for identifying and addressing performance problems when they occur. On occasion, this substandard performance may be the result of substance abuse. Supervisors, however, should not be expected to diagnose possible substance abuse problems. Instead, you can expect them to be able to identify the signs of poor job performance and follow standard company procedures for dealing with the employee.

The key to gaining effective supervisory support for your substance abuse program is to make sure all supervisors have been trained to understand the company's substance abuse policy and procedures, to identify and help resolve employee performance problems, and to know how to refer employees to available assistance so that any personal problems that may be affecting job performance can be addressed.

An effective training program will enable supervisors to do the following:

- Know the company's policy and understand their role in its implementation and maintenance
- Observe and document unsatisfactory job performance
- Confront workers about unsatisfactory job performance according to company procedures
- Understand the effects of substance abuse in the workplace
- Know how to refer an employee suspected of having a substance abuse problem to those who are qualified to make a specific diagnosis and to offer assistance

Again, supervision is one of the keys to an effective program in addressing the abuse of alcohol and other drugs in the workplace. It is important for the supervisor to focus on job performance and avoid acting the role of counselor or diagnostician. Supervisors should be able to recognize patterns of behavior that might indicate employee problems and be prepared to refer employees to the appropriate sources of help.

9.6 Educating employees

Educating all of your workers about substance abuse and your company's substance abuse program is a critical step in actually achieving the objectives of the program. Your substance abuse education and awareness program

SAMPLE ALCOHOL AND DRUG POLICY

XYZ Company is committed to providing a safe work environment and to fostering the well being and health of its employees. That commitment is jeopardized when any XYZ Company employee uses illegal drugs or alcohol on the job, comes to work with these substances present in his/her body, or possesses, distributes, or sells drugs in the workplace. XYZ Company has established the following policy with regard to alcohol and other drugs to ensure that we can meet our obligations to our employees, shareholders, customers, and the public.

The goal of this policy is to balance our respect for individuals with the need to maintain a safe, productive, and drug-free environment The intent of this policy is to offer a helping hand to those who need it, while sending a clear message that illegal drug use and alcohol abuse are incompatible with working at XYZ Company.

1. It is a violation of our policy for any employee to possess, sell, trade, or offer for sale illegal drugs or otherwise engage in the use of illegal drugs or alcohol on the job.

2. It is a violation of our policy for anyone to report to work under the influence of illegal drugs or alcohol (that is, with illegal drugs or alcohol in his/her body).

3. It is a violation of our policy for anyone to use prescription drugs illegally. (It is not a violation of our policy for an employee to use legally prescribed medications, but the employee should notify his/her supervisor if the prescribed medications will affect the employee's ability to perform his/her job safely).

4. Violations of this policy are subject to disciplinary action ranging from a letter of reprimand, to suspension from work without pay, up to and including dismissal.

It is the responsibility of supervisors to counsel employees whenever they see changes in performance that suggest that an employee has an alcohol or drug problem. Although it is not the supervisor's job to diagnose the employee problem, the supervisor should encourage such an employee to seek help and tell him/her about available resources for getting help. Because all employees are expected to be concerned about working in a safe environment, a supervisor also should encourage fellow employees who may have an alcohol or drug problem to seek help.

Figure 9.2

may differ from those of other companies, depending on your specific needs. However, a basic education program should achieve the following objectives:

*   Provide information about the dangers of alcohol and other drugs and how they can affect individuals and families.
*   Describe the impact that substance abuse can have on safety at work as well as the company's productivity, product quality, absenteeism, healthcare costs, accident rates, and the overall bottom line.
*   Explain in detail how your workplace policy applies to every employee of the company and the consequences for violations of the policy.

- Describe how the basic components of your overall program work, including the employee assistance program and drug/alcohol testing, if these are a part of the program.
- Explain how employees and their dependents, if included, can get help for their substance abuse problems, including how to access the company's employee assistance program or how to obtain services available from the community.

An effective education and awareness program is not a one-time effort. New hires should be informed immediately about your substance abuse policy and what is expected of them. Because of the regular turnover that many companies experience and the occasional changes and updates to the policy that may occur, education efforts need to be undertaken periodically. If your company is unionized, the union representatives can provide valuable assistance in the development and maintenance of an education and awareness program. Assistance in organizing and presenting this educational program can be secured from numerous external sources.

9.7 Employee assistance programs

Many employers are unsure whether they can or should offer or provide assistance to employees who have alcohol or other drug problems. Often they are concerned about the cost of providing assistance and the company's ability to continue to meet work demands while employees are receiving help.

Terminating employees with alcohol and other drug problems and then hiring a new worker may seem to be the most cost-effective approach. In some cases, starting fresh may be the best course of action. In most cases, however, it actually makes better sense — from a business as well as humanitarian point of view — to help employees overcome personal problems.

An employee assistance program (EAP) is a job-based program intended to assist workers whose job performance is being negatively affected by personal problems. Workers' personal problems may be caused by any number of factors, including substance abuse. Many employers have discovered that EAPs are cost-effective — they help reduce accidents, workers' compensation claims, absenteeism, and employee theft — and contribute to improved productivity and employee morale.

In this regard, one insured who had instituted a substance abuse prevention program during the previous policy year was asked about the acceptance of the program by the employees. He said their morale had increased and attributed it to the fact that they now had much more confidence in the "clear-headedness" of their coworkers. This was a commercial and industrial roofing company that encounters the hazards of working at heights and with hot tar.

If you are contemplating including employee assistance as part of your program, you may wish to contact other companies in your area that provide some type of employee assistance and learn about their programs — what they offer, how the service is provided, and the costs and results they are getting from the program. Also determine whether there is an EAP consortium available in your community that local businesses can join to receive EAP services at costs typically available only to larger companies.

In order for your EAP to be successful, your employees must view it as a confidential source of help. They must believe that they will not jeopardize their employment or future opportunities with the company by seeking help from the EAP. Conversely, they must also understand that the EAP will not shield them from disciplinary action for continued poor performance or violations of your company policy.

A company of almost any size can offer its employees the services of an EAP, which can be tailored to address the specific needs of your workforce. With a strong commitment from the management of the company, quality EAP professionals, and a clear understanding by all that employee assistance services do not offer "quick fixes," an EAP can be a valuable component of your comprehensive workplace substance abuse program.

Realize also that it is not always necessary to have a formal employee assistance program. For many companies, particularly small businesses, it is economically unrealistic to consider providing a formal program. However, the services that such a program provides are still available in a variety of ways — often within the budget of many small businesses. Some companies, for example, keep a folder with listings of resources available for employees. Regardless of how formal or informal you choose to be about providing such services, employee assistance can be a valuable component of your overall program.

9.8   Obtaining assistance

When developing a drug-testing program, there is no need to reinvent the wheel. Hundreds of thousands of companies have already implemented drug-testing programs that have proven effective. Furthermore, numerous independent and government agencies are ready and willing to offer information and assistance upon request.

Perhaps the most comprehensive source of information regarding workplace alcohol and drug programs is the National Clearinghouse for Alcohol and Drug Information (NCADI), a service of the Substance Abuse and Mental Health Services Administration (SAMHSA). This program provides numerous publications, video training programs, and posters, free or at a nominal charge.

Another good source of information and assistance regarding workplace alcohol and drug programs is insurance carriers. Since drug users account for more than twice as many workers' compensation claims as other employees, workers' compensation carriers are eager to offer assistance to any employer who expresses a willingness to implement a program intended to eliminate workplace alcohol and drug abuse. Furthermore, health insurance carriers are able to answer questions regarding the availability of coverage for alcohol- and drug-related problems for employees and their family members.

Other sources of assistance may include trade associations and other employers. Trade associations as well as community and business groups often offer materials on drug-testing programs. Furthermore, 75% of large businesses (1000 employees or more) are doing something to address alcohol and other drug abuse in their workplace, and many are willing to assist smaller businesses in addressing the problem.

For a more complete list of resources, including contact information, refer to Appendix E, National Resources for Alcohol and Drug Abuse Information.

9.9   Summary

The most important thing employers can do to help control the cost of substance abuse is to establish and enforce a policy that prohibits employees from using illegal drugs and abusing legal drugs or alcohol. Employers should train supervisors to monitor workers' job performance and to report any irregularities. However, supervisors should not be expected to diagnose possible medical conditions, such as substance abuse. Employers should also offer employees ongoing substance abuse education opportunities so that everyone in the workplace can be on the lookout for the hidden signs of substance abuse.

Hundreds of thousands of companies have implemented drug-testing programs. Since the individuals who abuse alcohol or other drugs are denied employment, are dismissed, or avoid companies who have such programs, they seek employment with companies that do not conduct drug testing. The result is that, as more and more companies adopt a drug testing policy, the companies that do not conduct testing are increasingly subject to encounter alcohol- and other drug-abusers as applicants and employees.

Employers must also realize that during the consideration and early stages of implementation of a substance abuse program, some employees will object, threaten to quit, or actually quit. A minimal amount of personnel turbulence will occur, but those who leave are either substance abusers or are so narrow-minded that they cannot see the benefits of a substance abuse program. Either way, that temporary employee turnover will soon result in a higher-quality work force.

Workplace substance abuse programs improve the workforce by eliminating those who abuse drugs and fail to rehabilitate and by deterring abusers from seeking employment with your organization. These programs directly affect your workers' compensation premium by securing credits, and they indirectly affect it by lessening the likelihood of injuries.

chapter ten

Combating workers' compensation fraud

Contents

10.1 Introduction

It is virtually impossible to turn on the television for any length of time without seeing an advertisement for an attorney trying to convince viewers that they are entitled to a disability settlement for virtually any work-related injury. The effect of this is that many people have begun to view workers' compensation as the modern-day lottery, and one that they have better odds of winning. Making workers' compensation fraud even more attractive is the relative ease by which it can be perpetrated without detection.

Although workers' compensation fraud can manifest itself in many different forms, evidence supports that the vast majority of workers' compensation claims are completely legitimate. Depending on the source, studies indicate that fraud is an element in 5% to 25% of workers' compensation claims. Using a conservative definition of workers' compensation fraud, it is likely that some element of workers' compensation fraud is involved in approximately 7 – 10% of submitted claims. Despite the relative low frequency, workers' compensation fraud is a serious problem that increases the cost of workers' compensation coverage, both for individual companies and for entire industries.

This chapter presents the effects of workers' compensation fraud; what factors increase the risk of workers' compensation fraud; the warning signs (or red flags) that should spark further investigation; and what employers can do to deter and detect workers' compensation fraud.

10.2   Effects of workers' compensation fraud

Estimates indicate that workers' compensation fraud costs industry an anywhere from 3.5 to 5 billion dollars each year. As with any business, the costs incurred by the provider of goods and services (in this case, workers' compensation insurance) is passed on to its customers. This means that fraud may cause insurance rates to increase for all companies, simply to recover the lost revenue of the insurance carriers.

Fraud may also cause premium increases to be shouldered by specific industries and by individual companies, based on past experience. To understand how fraud can cause premium increases for entire industries, it must be understood that the premium charged for each classification of worker is based on a statistical analysis of past claims submitted by all insurance carriers for each job classification. Hence, if a significant amount of claims is paid as a result of fraud among long-haul semitruck drivers, the rate for those drivers will increase proportionally. This affects all employers who have employees in that particular job classification.

In addition to affecting the cost of workers' compensation insurance for all companies who maintain workers' compensation coverage and for entire industries, workers' compensation fraud may increase premiums for individual companies. In fact, a single incident of workers' compensation fraud can have a significant impact on the future workers' compensation premiums paid by an individual company. With the cost of workers' compensation insurance coverage so substantial, unchecked workers' compensation fraud may threaten the very survival of a company. Regardless of how slight or dramatic, increased workers' compensation premiums restrict the financial resources of a business and inhibit the ability to expand or engage in other revenue-producing endeavors. Furthermore, increased costs for workers' compensation insurance hamper the ability of a business to operate competitively.

Insurance premiums can clearly be influenced by workers' compensation fraud, but the financial detriments to businesses do not end there. Workers' compensation fraud often causes lost productivity and the outlay of additional expenses of advertising, hiring, and training new employees, all of which eat into the profitability of a company. Consider an employee who has a legitimate work-related injury but exaggerates the severity of the injury to remain off work for an extended period of time. During the period that the claimant is off work, the company loses the productivity of that employee. If the company is unable to sustain the productivity level for the period that the claimant is off work, it may be forced to hire a replacement worker. The process of advertising, interviewing, selecting, hiring, and training a new employee not only costs the employer directly but also takes time from other employees and reduces their productivity.

The fraud also has an effect on the honest employees who remain at work. A number of different factors may cause tension and decreased morale to pervade the workplace. Understandably, these workers are frustrated when they see a coworker abusing the system at their expense because the coworker's absence translates into a more rigorous workload for them. If these employees understand that their employers' insurance premiums are affected by workers' compensation fraud, they realize that the cost of increased premiums may result in the absence of pay raises, reduced profit sharing, or even layoffs. Furthermore, because of the skepticism created by past incidences of workers' compensation fraud, the honest workers may feel distrusted by their employer or by the company's workers' compensation insurance carrier when submitting a claim for a legitimate work-related injury.

10.3   What constitutes workers' compensation fraud?

Although each state defines workers' compensation fraud somewhat differently, it is basically defined as knowingly making false or misleading statements or actions with the intent to profit financially. Workers' compensation fraud typically conjures thoughts of a claimant who alleges an injury that is not legitimate, or a claimant who exaggerates the severity of a claim to increase the period of time away from work or the amount of a potential financial settlement. The reality is that workers' compensation fraud can be far more broadly defined. As such, workers' compensation fraud can be viewed as an offense perpetrated not only by claimants but also by corrupt physicians, attorneys, and even employers and insurance carriers themselves.

Although workers' compensation fraud is defined differently by each state's legislation, a list of acts likely to be considered fraudulent are described below. The following discussion, although not exhaustive, demonstrates that workers' compensation fraud is not limited to the blatant fabrication of a claim.

10.3.1   Claimants

Probably the most blatant form of workers' compensation fraud involves a claimant who fabricates an injury and submits it as a workers' compensation claim. In such cases, the claimant does not incur an injury at all but is aware that if he or she can convince the physician that an injury exists or can find a physician with whom to collude, time off work and indemnity benefits or even a disability settlement can be obtained. As with most forms of workers' compensation fraud, this form of fraud often involves the claim of a soft-tissue injury such as a strained back or sore wrist because subjective claims of pain and lack of mobility are medically difficult to confirm or deny.

Similar to the above form of workers' compensation fraud, one of the most common forms of workers' compensation fraud occurs when a claimant submits a workers' compensation claim for an injury that was not work-related in an attempt to have it paid by the company's workers' compensation carrier as opposed to personal health insurance. In contrast to the previous form of fraud, this involves an individual who actually sustained an injury. In the event that the claimant has no health insurance, this type of fraud is often committed to avoid out-of-pocket medical expenses. However, individuals who have health insurance may also do this to avoid paying deductibles, as workers' compensation does not involve employee-paid deductibles.

Another type of fraud perpetrated by a claimant involves an individual who files a workers' compensation claim with two or more employers. This is appropriately dubbed "double-dipping." A twist on the double dipping type of fraud occurs when a claimant draws unemployment benefits concurrently with workers' compensation indemnity benefits and does not advise either entity of the benefits drawn from the other. This may also constitute another type of fraud, as the claimant is defrauding not only the insurance carrier but also the state unemployment insurance program.

Another very common type of fraud is a claimant who exaggerates the extent of a legitimate injury to sustain indemnity (wage replacement) benefits. This is likely one of the most frequent forms of workers' compensation fraud and is particularly prevalent with claimants who allege soft-tissue injuries. This form of workers' compensation fraud often manifests itself in malingering, the term used to describe a claimant who remains off work beyond the time that is medically necessary, commonly espousing questionable claims of pain or lack of mobility. In a similar type of fraud, the claimant exaggerates the extent of a legitimate injury to receive an increased financial settlement. With this variation of fraud, the claimant's goal is to convince the medical provider, the insurance carrier, and the employer that the alleged injury has resulted in a permanent physical disability.

Although with each of the above versions of fraud the claimant is the perpetrator, an unusual twist involves indemnity benefit checks that continue to be cashed after the claimant has died. The claimant can hardly be held accountable for this type of workers' compensation fraud. Instead, the perpetrator of this type of fraud is the person cashing the checks.

Yet another form of fraud occurs when a claimant continues to receive workers' compensation indemnity benefits while employed by a different company. This is a relatively common form of fraud and involves a claimant who is receiving indemnity benefits because of his alleged inability to perform work. While receiving these benefits the claimant is employed by another entity and fails to declare that income to the workers' compensation carrier providing the indemnity benefits. In an attempt to avoid detection, individuals who commit this type of workers' compensation fraud frequently obtain employment through which they are paid cash.

10.3.2   Medical providers

Workers' compensation fraud is not only perpetrated by the people who sustain or allege injuries but can also reveal itself in the actions of the medical community. One such form of fraud involves a medical provider who inflates the amount of bills for medical services. Since insurance carriers generally pay medical claims based on a predetermined fee schedule, inflated billings may more cleverly be facilitated by performing unnecessary and expensive diagnostic testing or by billing for services that were not rendered.

Another type of workers' compensation fraud perpetrated by medical providers occurs when a medical provider makes unwarranted referrals to other medical providers. These other medical providers may be owned by the medical provider referring the patient, or the medical provider making the referral may get a kickback from the entity to which the referral is made. A similar type of fraud involves a conspiratorial relationship between the medical provider and the claimant. With this type of fraud a medical provider and a claimant conspire to exaggerate the severity of an injury. The injury that the claimant sustained is generally legitimate and work-related but is exaggerated to necessitate additional treatments or diagnostic testing. Likewise, a medical provider may knowingly falsely diagnose a claimant's injury. When this occurs, the claimant and the medical provider may have conspired, or the medical provider may provide the false diagnosis purely for personal financial gain, without the claimant's knowledge. A medical provider may also falsely indicate that an injury is work-related to ensure payment for medical treatment. Physicians know that they will receive payment for workers' compensation claims. Hence, in situations in which the claimant is indigent and does not maintain other forms of health insurance coverage, the medical provider may diagnose an injury to be work-related when there is no evidence to support such a diagnosis.

Similarly, a medical provider who falsely indicates that a preexisting medical condition was caused by recent work activity is also committing workers' compensation fraud. In many situations, a preexisting medical injury is not a compensable injury, even if aggravated by recent work activity.

Hence, the issue of injury causation is pivotal. With this type of fraud, the medical provider may attempt to conceal or underplay the severity of a preexisting medical condition to ensure that payment of medical expenses will be covered through workers' compensation.

10.3.3 Legal professionals

Many things contribute to the negative perception of the legal profession that some people hold. However, nothing is more repulsive than legal professionals who use their position of authority and trust to commit fraud, without regard for how their actions may affect others.

One of the most blatant forms of fraud perpetrated by legal professionals occurs when a legal professional solicits individuals to submit claims. When this type of fraud is perpetrated, the legal professional often finds likely accomplices outside unemployment offices and convinces them to submit workers' compensation claims for false injuries, which are alleged to have occurred when the claimant was last employed.

Another type of fraud perpetrated by legal professionals involves a legal professional who instructs his client (the claimant) to mask signs of recovery. This is done in an attempt to lengthen the period during which indemnity benefits are received or to increase the amount of a settlement.

Just as medical providers commit fraud when they illegally conspire with other providers of medical or rehabilitative services, a legal professional commits fraud when he or she directs claimants to conspiring medical providers. In these situations, the legal professional may either be getting a kickback from the clinic or be conspiring with the medical provider to render a specific diagnosis.

Although virtually impossible to prove, a workers' compensation judge or arbitrator who finds in favor of a claimant simply to ensure that a personal friend (attorney) receives a portion of the settlement is committing fraud. Whereas this type of fraud may not be prevalent, it is provided as an example to demonstrate that nobody within the workers' compensation system is incapable of committing fraud

10.3.4 Employers

Whereas employers are often one of the key parties negatively affected by workers' compensation fraud, they are capable of committing workers' compensation fraud. One example is an employer who, in response to an employee's injury claim, knowingly and falsely claims that no injury occurred or that the injury was not work-related. Employers may be motivated to do this in an attempt to manipulate their workers' compensation loss history by directing employees to submit legitimate work-related injury claims through personal or group health insurance.

Another type of fraud involves an employer who manipulates classification and payroll data. Although this is a different realm of workers' compensation fraud, employers who abuse the system to pay a reduced premium may be committing what is commonly termed "premium fraud." This includes submitting deflated payroll information during an audit, misrepresenting the type of work performed by employees to have them classified in a position with a lower rate, and unscrupulously manipulating the experience rating system by failing to report claims.

An extreme version of premium fraud involves an employer who changes the name of the company to conceal a poor loss history. Because the loss history of a company has such a significant impact on the premium, a company owner might create a new company but continue the same type of work and retain the same staff in an attempt to elude a poor loss history.

10.3.5   Insurance carriers or third-party administrators

Insurance carriers themselves are capable of committing workers' compensation fraud. One example is an insurance carrier that denies a claim known to be compensable under current workers' compensation legislation. Fraud is also committed when a fictitious insurance carrier, posing as a state-approved insurer, collects premiums. These bogus companies produce counterfeit certificates of insurance to trick their victims. Unfortunately, these fictitious companies dupe far too many small employers into surrendering substantial payments before they disappear, never responding to injury claims.

10.4   Factors that affect the potential for fraud

The prevalence of workers' compensation fraud stems, at least in part, from the relative ease of abusing the system. Because workers' compensation is a no-fault insurance (meaning that benefits are paid regardless of who is at fault) it is by its very nature a system that has a high potential for abuse. However, there are factors that influence the potential for the workers' compensation system to be abused. Several of these factors are presented below.

10.4.1   Management's interaction with employees

One factor that could increase the potential for workers' compensation fraud is the manner in which the employer treats employees who report work-related injuries. Treating legitimate claims as fraudulent can increase the potential for workers' compensation fraud by creating an adversarial relationship between company management and other employees. This adversarial attitude between managers and subordinate workers can spread throughout the entire organization. Consider the business concept that a satisfied customer will tell another person of his experience with the company with which he has had a good experience, whereas a dissatisfied customer will likely tell ten times as many people of his poor experience.

Similar to the manner in which the employer treats employees who report work-related injuries is the relationship between management and workers in the absence of workers' compensation claims. A good management/employee relationship fosters a sense of allegiance to the employer (or at least to the manager). A solid and favorable relationship between management and employees also increases an employee's sense of job satisfaction. Whereas allegiance to the employer and a sense of job satisfaction serve to deter such detrimental acts as fraud, the opposite is the likely result of poor management/employee relationships.

10.4.2   Supervision of employees

Closely related to the type of interaction between management and employees is the amount of interaction. Some businesses, by their very nature, employ individuals of whom there is little or no direct supervision and/or coworker interaction. Examples are outside sales employees and delivery drivers. These employees can very easily (and fraudulently) report an injury that happened away from work as a workers' compensation claim because there are plenty of times that no witnesses would be expected to have seen the injury-producing incident. Although there may be no feasible means of providing greater supervision of some workers, employers should be aware that the lack of supervision might increase the potential for workers' compensation fraud.

10.4.3   Administrative efforts

The lack of administrative efforts to counter abuses of the workers' compensation system is another factor that may increase the potential for workers' compensation fraud. Far too many employers fail to enact any proactive measures to deter workers' compensation fraud and seemingly cross their fingers, hoping that they will not become victims. Likewise, the absence of employee training relative to workers' compensation and workers' compensation fraud may increase the potential for fraud. Whether it is ignorance on the part of the management or the line employees, a lack of knowledge of how workers' compensation insurance works can create an atmosphere in which fraud is neither deterred nor detected. Furthermore, if employees or managers think that the insurance carrier pays for workers' compensation claims without any financial consequences for the employer, even employees who have an allegiance to their employer may be tempted to abuse the system. Hence, all employees must understand how fraud affects them individually and how it affects their employer.

Regardless of the actions or inaction of management that first contribute to workers' compensation fraud, a lack of effective means of detecting and stopping fraud once it occurs can itself increase the potential for future system abuses. In short, the past success of one or more employees in abusing the workers' compensation system may encourage other employees to view workers' compensation fraud as presenting little risk. This dangerous situation may cause abuse of the workers' compensation system to run rampant throughout an organization.

10.4.4   Employee benefits

Although not traditionally considered to be a countermeasure to workers' compensation fraud, employee benefits such as healthcare coverage, sick days, and vacations have the potential to influence workers' compensation fraud. Consider an employee who is not provided affordable healthcare coverage through his employer. If such a person is injured away from work, he or she is faced with the decision to pay for medical expenses via out-of-pocket expense or to wait until he or she returns to work to try to pass it off as a work-related injury. If personal finances are limited, even a relatively honest employee can be tempted to submit a fraudulent claim.

Even if an employee has adequate healthcare coverage, that employee may still be tempted to submit a fraudulent workers' compensation claim if he or she has no sick leave or vacation benefits. In such circumstances an employee may view workers' compensation as a means of attaining benefits to which he or she feels entitled. Furthermore, employees who face time away from work as a result of a non- work-related injury may be motivated to use workers' compensation as a means of obtaining wage replacement benefits to which he or she would not otherwise be entitled.

10.4.5   Too much or too little work

Both too much and too little work can contribute to the potential for workers' compensation fraud. Employees who are overworked and required to work long days or who are required to work on their normal days off may view workers' compensation as a means of getting time off work to relax. Conversely, employees who have too little work may conjure thoughts of being laid off or down-sized out of a job. Whether the fear is real or imagined, submitting a fraudulent workers' compensation claim may appeal to some as a means of sustaining wages for a period longer than unemployment benefits would offer.

10.4.6   Regulatory constraints

Despite an employer's effort to deter, detect, and properly respond to workers' compensation fraud, their hands remain tied to some extent by regulatory constraints. The most evident of these is the Americans with Disabilities Act (ADA). Prior to the advent of the ADA, employers were able to ask job applicants if they were currently receiving workers' compensation benefits or if they had ever submitted a workers' compensation claim. However, asking these questions now may constitute a violation of the ADA and some state laws.

10.5   Warning signs (red flags)

The auditors of the Internal Revenue Service use a series of "red flags" that, while not necessarily indicating a particular taxpayer has committed income tax evasion, sparks closer scrutiny to confirm or negate that possibility. Similarly, as used when referring to workers' compensation fraud, "red flags" are indicators that point to the need for further investigation of a claim to determine its legitimacy. Of course, the applicability of one or even several "red flag" indicators does not necessarily indicate the existence of workers' compensation fraud.

Although not exhaustive, the following is a list of warning signs that a workers' compensation claim should be closely scrutinized. The signs can be divided into three separate groups: red flags concerning the injury, red flags concerning the claimant, and red flags concerning medical treatment.

10.5.1   Red flags concerning the injury

10.5.1.1   Delay in reporting an injury

When there is an unexplainable delay in reporting an injury, it is sometimes a result of the claimant sustaining an injury that initially seems very minor but graduates to more severe symptoms over time. However, an unexplainable delay in the reporting of an injury may also be an intentional attempt to make it difficult for the employer to question the circumstances surrounding the alleged injury-producing incident. For example, a claimant who alleges a back injury from lifting items of stock last week knows that it is difficult for an employer to confirm or deny the claim based on physical evidence or witnesses. It is notable that this is not only a red flag indicator of possible workers' compensation fraud but may also be justification for the insurance carrier to deny a claim.

10.5.1.2   No witnesses

When there are no witnesses to an alleged injury-producing incident, the ability of the employer or workers' compensation carrier to contradict the claimant's version of an injury is obviously hindered. However, many work-related injuries do occur without witnesses. Therefore, the applicability of this red flag indicator should not cause undue alarm.

10.5.1.3   Insufficient detail

When the details of an injury-producing incident are few or ambiguous, the employer should be aware of the possibility that the injury was fabricated. A vague description of an injury-producing incident may indicate that the employee did not expect the employer to question the alleged circumstances surrounding the injury and therefore has not thoroughly contrived a believable story.

10.5.1.4 Injury is inconceivable

It is not unreasonable for employers to use their own knowledge of the work performed by individuals to question the legitimacy of a claim. Hence, if the claimant's description of an injury appears implausible considering the type of work that the employee performs, it may warrant further investigation. Although an extreme example, it may be implausible for a clerical employee to get injured while assisting loading a truck. A less extreme example is an employee who claims an injury from lifting an object that is almost always lifted with a forklift.

10.5.1.5 Soft-tissue injuries

Many work-related injuries are not visible, such as a strained back, carpal tunnel syndrome, and sprains. These are often termed "soft-tissue injuries." These types of injuries are not only difficult to confirm or deny, but they rely on the claimant's subjective indications of pain and range of motion to determine when the claimant is able to return to work. For this reason, these types of injuries may warrant further investigation to confirm or deny their legitimacy.

10.5.1.6 Inconsistencies between injury and incident

If an incident allegedly resulted in a significantly more serious injury than would reasonably be expected, the claimant may be exaggerating the extent of the injury to obtain additional time off work or receive a permanent disability payment. For this reason, an injury that seems significantly more serious than would be expected, based on the claimant's description of the injury, would constitute a red flag. For example, a claimant strained an arm reaching to retrieve a 15-pound box and is alleging a lack of mobility so great that it precludes the performance of job duties.

10.5.1.7 Allegations of fraud

Although rumors are undesirable in any workplace, rumors circulating that an injury is not legitimate should spark further investigation. Unlike many other workplace rumors, there is often validity to rumors relating to the legitimacy of workers' compensation claims. Perhaps this is because claimants may confide in their coworkers or may have previously informed coworkers of an injury or chronic pain that is later submitted as a workers' compensation claim.

10.5.1.8 Monday morning injuries

One of the most common red flags of workers' compensation fraud is an injury that reportedly occurred early on a Monday or the first day of work after one or more days off. An alleged injury submitted on a Monday morning may involve an injury that was sustained while the employee was off work over the weekend, but the injured employee waited until returning to work to submit it as a workers' compensation claim.

10.5.1.9   Recent purchase of disability insurance

The employer may be justified in investigating a claim further if the claimant has recently purchased disability insurance that will pay benefits concurrently with workers' compensation benefits. Although obtaining an insurance policy that pays concurrently with workers' compensation is a legitimate practice, a claimant may be motivated to submit a fraudulent claim if he or she can draw a greater income from the combined benefits of workers' compensation and disability insurance than would otherwise be netted from pure income.

10.5.2   Red flags concerning the claimant

10.5.2.1   New employee

Although it is true that new employees are statistically more likely to sustain a work-related injury, employers should still be wary of a claimant who is a recently hired employee. Sad but true, some individuals view the workers' compensation system as an opportunity to get rich quick. There is little benefit for these employees to remain employees for years before attempting to abuse the system.

10.5.2.2   No health insurance coverage

As mentioned previously, employees who do not have health insurance may seek to have an injury or illness which is not work-related treated as a workers' compensation claim because they cannot afford, or do not want to pay the out-of-pocket expense for medical treatment. For this reason, prudent employers should scrutinize claims submitted by employees without health insurance coverage, unless it is clearly obvious that the injury occurred at the workplace.

10.5.2.3   No remaining time off work

An employee who is unable to take time off work legitimately may view workers' compensation as a way to attain paid time off. Therefore, an employee who has used all of his or her accrued vacation and sick days and then submits a workers' compensation claim may warrant a closer look.

10.5.2.4   Personal financial problems

With workers' compensation viewed as the modern-day lottery by some unscrupulous individuals, the potential to score a financial settlement to alleviate personal financial problems is a strong motivation for some individuals to commit workers' compensation fraud. Supervisors and other employees may volunteer information about a claimant's personal financial problems simply by demonstrating a concern for the claimant. Just as any single red flag is likely insufficient reason to leap to any conclusions about the legitimacy of a claim, a claimant's poor personal financial status is not alone indicative of fraud. If this were the case, most of the workforce would be suspect with every claim.

10.5.2.5   *Physically active outside work*

If a claimant is physically active outside work, he or she is at greater risk of incurring an injury away from work than his or her less active counterparts. For example, an employee involved in organized sports may be injured while participating in a sporting event and submit that injury as a workers' compensation claim with the employer.

10.5.2.6   *Injury repeater*

If an employee has a history of alleging work-related injuries, particularly minor injuries, soft-tissue injuries, or injuries that have resulted in time off work, the employer should be aware that this may not simply represent an employee who has not mastered the skill of performing his or her job safely but someone committing workers' fraud.

10.5.2.7   *Ties to other claimants*

A claimant who has close friends or relatives who have submitted workers' compensation claims in the past that have resulted in indemnity benefits or settlements can be a red flag. More frequently than many people outside the workers' compensation insurance industry may believe, workers' compensation fraud is contagious among friends and relatives. Individuals who see their friends or relatives committing fraud without detection are more likely to try it themselves. They may even get advice from those who have been successful in the past. Similarly, individuals who have submitted legitimate claims in the past may unknowingly provide the motivation for a friend or relative to seek similar benefits through illegitimate means.

10.5.2.8   *Inconsistencies from the claimant*

Descriptions of an injury or incident that are inconsistent with what was originally reported should be viewed as red flags. Just as prosecutors of criminal cases uncover inconsistencies in subsequent statements made by defendants, inconsistent statements about the alleged injury-producing incident or the injury itself can suggest fraudulent workers' compensation claims.

10.5.2.9   *Unusual familiarity with workers' compensation*

A claimant who is unusually familiar with the workers' compensation system and workers' compensation legislation may indicate someone who has had experience in submitting claims and receiving benefits or who has "done his homework" about workers' compensation, so that he or she knows how to abuse the system without detection.

10.5.2.10   Objections to management controls

Throughout this text, management controls are presented to address workers' compensation costs and the abuse of workers' compensation. A red flag should be raised if an employee resists these controls. For example, a claimant who objects to management visiting him or her at home may indicate that the claimant is frequently not at home due to another job or engages in activity at home which he or she is allegedly unable to perform. Similarly, opposing modified duty assignments may be a possible indicator of fraud. However, the employer should be aware that failure to accept an appropriate modified duty assignment might itself constitute justification for the denial of indemnity benefits.

10.5.2.11   Does not provide street address

Another red flag of potential fraud is if the only home address that the company has for the claimant is a post office box. Claimants who commit fraud realize that this significantly hinders the ability of company management to visit the claimant at his or her home or to conduct surveillance.

10.5.2.12   Unable to contact claimant

If while off work as a result of a workers' compensation claim, a claimant is frequently unable to be contacted by telephone at his or her home or can be contacted only through a friend or relative, the employee is possibly working elsewhere while receiving workers' compensation indemnity benefits.

10.5.2.13   Indemnity checks sent to residence

Another red flag is if a claimant specifies that indemnity benefit checks be sent to his or her home. The employer should be aware that this might be done to avoid contact with company management. This is not only a red flag of possible workers' compensation fraud but is a hindrance for the employer who is trying to maintain good relations with the claimant through routine contact.

10.5.2.14   Disgruntled or uncooperative attitude

A disgruntled or uncooperative attitude may appear either before or after a claimant reports an injury. Regardless of when this attitude becomes evident, it should be considered an indicator of the potential for fraud. As previously indicated, poor employee/employer relations contribute to the incidence of fraud. A disgruntled or uncooperative employee may be indicative of those poor relationships.

10.5.2.15   Observed performing work elsewhere

If it has been reported that the claimant has been seen working for another employer or engaged in work that he or she is believed to be unable to perform, further investigation is always warranted. This is one of the strongest indicators of the possibility of workers' compensation fraud.

10.5.2.16 Rapid retention of legal counsel

Another red flag indicator of workers' compensation fraud is if the claimant obtains legal representation soon after submitting a workers' compensation claim. The workers' compensation process is designed to be effective without the need of attorneys. As such, the involvement of an attorney at the onset of a claim is a strong indicator of the potential for a claim to be fraudulent.

10.5.2.17 Reputation of legal counsel

Whereas every claim that involves an attorney is not necessarily suspect of fraud, if an employee has obtained an attorney with a reputation of litigating questionable claims, it should be viewed as a red flag. Although some of these attorneys obtain their notoriety from television commercials touting workers' compensation as a specialty, some less-known law firms may have equally questionable business practices. Workers' compensation carriers are likely familiar with the attorneys whose involvement is a red flag.

10.5.2.18 Seeks rapid settlement

Another strong red flag indicator of fraud is if a claimant or the claimant's attorney insists on a rapid financial settlement. This is sometimes done with the hope that the insurance carrier will take the bait and not look into the claim with as much scrutiny as it may warrant.

10.5.2.19 Subject to termination of employment

If a claimant is the subject of an imminent or suspected termination or layoff or is near the end of a probationary period, further investigation is warranted. Although an employee who is terminated or laid off may be eligible for unemployment benefits, a workers' compensation claim offers the potential for indemnity benefits which can far exceed the maximum received from unemployment (due to length of period for benefits received and the potential for a lump-sum settlement). This is a particular concern for older workers as they may have more difficulty obtaining other employment.

10.5.3 Red flags concerning medical treatment

10.5.3.1 Contradictory evaluations

It is not uncommon for a claimant to be treated by more than one physician during the treatment period. However, a medical examination that contradicts the original diagnosis is definitely a red flag for potential fraud. It might indicate that the initial medial treatment provider is corroborating with the claimant to exaggerate the severity of a claim. It may also indicate the claimant's ability to influence one physician's diagnosis with subjective claims of pain or restricted mobility, whereas the contradicting diagnoses from subsequent medical treatment providers are relying on more objective diagnostic tools.

10.5.3.2 *Reputation of medical provider*
Just as the questionable reputation of an attorney warrants further investigation, a claimant who is seeking medical treatment from a medical provider with a reputation of questionable diagnosis and treatment practices is also a red flag. Although managed care programs try to weed out physicians with questionable ethical practices, the reality is that there are and will always be some physicians who provide diagnoses and treatments that are solely intended to increase their income.

10.5.3.3 *Missed medical visits*
An employee who misses physician visits and/or rehabilitation treatments may indicate that the employee knows that the treatments are unnecessary. Although there are completely legitimate reasons for missing scheduled visits, missed medical visits should be viewed as a red flag for workers' compensation fraud.

10.5.3.4 *Treatment is inconsistent with injury*
If treatment for a work-related injury appears more intensive than the alleged injury seemed to require, it may indicate a medical provider who has prescribed a treatment plan intended to inflate the bill for services. Whereas most employers do not have a background in the diagnosis and treatment of injuries, intuition plays a factor, particularly when expensive diagnostic procedures are billed for relatively small injuries.

10.5.3.5 *Medical referrals to neighboring practice*
Referrals to a separate medical provider under common ownership or in close proximity to the referring medical provider can be a completely legitimate practice. However, they should cause closer scrutiny. The referral of a claimant to a separate medical provider under common ownership creates an obvious red flag, and the referral of a claimant to a medical treatment provider within close proximity to the referring physician may indicate that the referring physician has a nonprofessional relationship with a nearby medical practice.

10.5.3.6 *Doctor shopping*
Fraud may be revealed if a claimant changes medical providers, particularly if the initial physician has released the claimant to return to work. This red flag may reveal what is commonly known as "doctor shopping," wherein the claimant seeks treatment from a number of different physicians until one is found that will diagnose time off work or a permanent disability.

10.6   Investigating suspect fraudulent claims

Although insurance companies investigate claims that they believe to be fraudulent, it is not cost-effective for insurance companies to send a staff investigator or contract a private investigator to collect videotape surveillance of all suspect claims. If this were commonly done, premiums would surely increase to a prohibitive level. For this reason, it is incumbent on employers to conduct at least preliminary investigations of suspect claims and provide any information supporting or negating the suspicion of fraud to their insurance carrier.

To provide the insurance carrier with the ammunition needed to fight a suspect claim, the employer must consider both the alleged loss event and the attributes of the claimant. The insurance carrier or its contracted investigators cannot obtain much of this information.

If it can be done safely, employers should re-enact the alleged injury-producing incident to determine if the alleged injury could have been sustained in the manner that the claimant described. Although a collateral benefit of this process, the re-enactment of injury-producing incidents also enables the employer to gain a better understanding of the incident, so that policies or procedures can be implemented to prevent similar occurrences.

Beyond re-enacting the alleged injury-producing incident, the employer should consider the claimant's coworkers as a potential source of useful information. Coworkers may have seen something that can either confirm the claim or provide further justification to question it. Coworkers may have heard the claimant complain of an injury unrelated to work or may know of personal financial problems, the claimant's impending voluntary termination, or relatives of the claimant who are receiving benefits from workers' compensation. All of this information is useful in determining if a claim should be considered suspect.

As a more comprehensive technique, employers may wish to create a checklist from the above list of red flags and provide that checklist to the workers' compensation carrier with each claim that is not clearly legitimate. Figure 10.1 is a sample workers' compensation fraud red flag checklist.

10.7   Methods of deterring and detecting fraud

Insurance companies use highly trained auditors, fraud investigators, and specialized computer software to aid in the detection of workers' compensation fraud. However, because of familiarity with the claimant and the work performed by the claimant, employers are in a unique position to detect and deter workers' compensation fraud within their company.

Employers who are serious about curbing fraud must realize that individuals who try to abuse the workers' compensation system do so when employed by a company that doesn't have effective controls to counter workers' compensation fraud. For this reason alone, it is incumbent on responsible employers to adopt policies that are intended to deter workers'

	Workers' Compensation Fraud **Red Flag Checklist**

Claimant's Name	Date of Injury

Mark applicable red flag indicators. Describe applicable red flag indicators on reverse.

NOTE: "Red flags" are indicators that indicate the need for further investigation of a claim to determine its legitimacy. Therefore, the applicability of one or even several "red flag" indicators is not necessarily indicative of the existence of workers' compensation fraud.

	There was a unexplainable delay in reporting
	There were no witnesses to the alleged injury-producing incident
	Insufficient detail was provided surrounding the injury-producing incident
	The alleged injury seems inconceivable considering the work which the claimant performs
	The injury is not visible (e.g., soft tissue injury)
	The degree of injury is not likely to result from alleged injury-producing incident.
	There have been allegations or rumors of fraud and/or the claimant has been observed working elsewhere
	The incident was reported on a Monday morning (or after one or more days off work)
	The claimant has recently purchased disability insurance
	The claimant is a new employee
	The claimant has no health insurance coverage
	The claimant has used all available sick days and vacation days
	The claimant is known to have personal financial problems
	The claimant is physically active outside
	The claimant has submitted workers compensation claims in the past
	Inconsistencies have been revealed from the claimant's initial description of the injury-producing incident
	The claimant is unusually familiar with the workers' compensation system
	The claimant is uncooperative and/or objects to administrative controls intended to address workers' compensation fraud
	The claimant does not provide a street address for a residence
	The employer is frequently unable to contact the claimant while off work due to an alleged injury
	The claimant obtained legal representation soon after the alleged incident and/or has obtained legal counsel with a questionable reputation
	The claimant has indemnity checks mailed to his/her residence
	Subsequent medical evaluations apparently contradict the initial evaluation
	The employee has missed scheduled physician visits or rehabilitation appointments
	The treatment being provided seems more extensive than the injury warrants
	The claimant has changed medical providers more than once after the initial treatment
	The claimant has been referred to a medical provider close in proximity to the referring medical provider

Figure 10.1

compensation fraud and to detect it when it occurs. The following are measures that employers can undertake to both deter and detect workers' compensation fraud.

10.7.1 Selecting the right workers' compensation carrier

One of the first things that an employer can do to counter workers' compensation fraud is to select a workers' compensation provider that has a strong stance against fraud. Employers should be aware that some insurance companies seemingly fail to treat workers' compensation fraud with the fervent repugnance that it deserves. Many employers who have been insured with a variety of workers' compensation providers in the past have had at least one bad experience in which ample evidence appeared to be present to deny or at least ardently investigate a claim, but despite the evidence of fraud the insurance carrier was unresponsive to the employer's appeals. Whereas insurance carriers will rarely admit indifference to workers' compensation fraud when soliciting a company's business, it doesn't take too long before insurance carriers get a reputation among employers, insurance agents, insurance brokers, and insurance consultants. For this reason employers should ask their prospective insurance carriers, insurance agents, and even other companies who are or have been insured by the prospective carrier of the insurance carrier's stance on workers' compensation fraud and the level of proof necessary to spark further investigation.

10.7.2   Selecting the right physicians

Just as it is important to select the right workers' compensation carrier, it is also important for employers to use reputable physicians to provide the initial treatment and diagnosis of work-related injuries or to counter the questionable diagnosis and/or treatment of less reputable physicians. Reputable physicians may also be able to identify inconsistencies in the claimant's subjective description of pain or mobility. Many states permit employers to require claimants to seek treatment from preselected physicians. Although effective, there are legislatively imposed restrictions regarding this practice of which the employer must be aware.

10.7.3   Designating a workers' compensation claims coordinator

Employers can aid in the detection and deterrence of workers' compensation fraud through selecting a responsible management employee to be the company's designated workers' compensation claims coordinator. This employee, among other duties, should be given the responsibility of noting the warning signs of workers' compensation fraud. Furthermore, the workers' compensation claims coordinator should be charged with educating employees of the consequences of workers' compensation fraud and maintaining effective communication with the insurance carrier, claimants, and medical treatment providers.

10.7.4   Implementing effective administrative efforts

One effective administrative effort, which has the effect of both deterring and detecting workers' compensation fraud, is a modified duty program. Such programs enable the claimant to work within their prescribed physical abilities as indicated by their treating physician. Often fraudulent filers of workers' compensation claims will remain off work as long as possible and then quit once they have been released to return to work without physical restrictions. Modified duty provides an opportunity for these workers to return to work even if they are unable to perform their normal job duties. However, many fraudulent filers will voluntarily terminate their employment rather than return to work.

Employers can also help deter incidents of workers' compensation fraud by providing employees with affordable healthcare coverage. It stands to reason that an employee who has healthcare coverage will be less motivated to submit a workers' compensation claim for an injury that is not work-related than an employee without such coverage.

Another effective administrative tool is to routinely (e.g., weekly) obtain signed written statements from claimants who are off work due to a workers' compensation claim indicating that they have not engaged in any type of employment during the previous week. As with any administrative control, this practice should be administered consistently and not just for claims suspected of fraudulent activity.

10.7.5   Exercising caution in hiring and firing

As a means of deterring workers' compensation fraud, employers should verify information on job applications. A person who is willing to lie on an application to obtain a job will likely have no problem with lying to receive workers' compensation benefits. This should include comparing the Social Security number provided on the job application with the applicant's Social Security card, as most workers' compensation carriers use the claimant's Social Security number as a means of tracking past claims.

Beyond verifying the accuracy of the information supplied on job applications, employers should review the background of prospective employees and obtain as many details from previous employers as possible. As previously mentioned, the Americans with Disabilities Act (ADA) restricts some of the previously used methods of inquiry. Therefore, it may be necessary for employers to consult with an attorney who is familiar with employment law to determine what is acceptable and unacceptable under current legislation.

Once the information on the job application has been verified and background information about the job applicant has been obtained, the employer should consider conducting drug testing of new-hire employees. By definition, illicit drug use is a violation of criminal law. An individual who violates laws relating to illicit drug use may lack the moral character to refrain from workers' compensation fraud.

Frequently done at the same time as new-hire drug testing, employers should consider requiring employees to submit to physical examinations on initial employment. Again, because of the Americans with Disabilities Act, it is no longer acceptable for employers to ask preemployment questions regarding past injuries. Instead, employers' questions regarding past injuries or current disabilities are limited to asking the applicant if he or she has any physical impairments that would preclude the performance of defined job duties. Whereas the findings of a physical examination on initial employ-ment may not preclude the applicant from being hired, provide baseline data that is helpful in determining the level of fitness at the time of hire. This is important, as following a work-related injury the employer is responsible only for the portion of disability incurred as a result of that work-related injury. For example, an applicant with a 5% disability of his back who is hired may injure his back while at work, resulting in a 10% disability of his back. The employer's responsibility is for 5% of the disability, not the entire 10%. However, without a physical examination, the employer and its insur-ance carrier may never know the level of fitness of the employee on hire.

Because workers' compensation fraud is prevalent among employees who are terminated or are laid-off, it is wise for employers to conduct exit interviews with employees that include documentation signed by the employee that he or she has not sustained a work-related injury that has not yet been reported.

10.7.6   Educating employees

Once due diligence has been exercised in selecting employees, the employer should embark on a campaign to educate employees about workers' com-pensation fraud. As stated previously, employers who are perceived as hav-ing insufficient controls to protect against workers' compensation fraud are the employers who are the mostly vulnerable victims. Although an employer may have mechanisms to identify fraud when it occurs, the value of those controls is diminished if the employees are not familiar that they exist. It is therefore incumbent on employers to educate employees.

The most obvious starting point for educating employees is to define workers' compensation fraud. Employers may wish to provide examples, as outlined in this chapter, to demonstrate that company management is aware of the various forms in which workers' compensation fraud can manifest itself. In doing so, employees are made aware that the company management knows the schemes perpetrated by unscrupulous individuals and that man-agement has devised means to detect fraud when it occurs.

Beyond defining workers' compensation fraud, employers should make clear the civil and criminal penalties of workers' compensation fraud. Going one step beyond educating employees verbally, employers should post signs describing the potential penalties for individuals who commit compensation fraud. As a part of this education process, employees should be aware that the company will seek civil remedies for the repayment of benefits and will pursue criminal prosecution for all cases of workers' compensation fraud.

With the knowledge of how workers' compensation fraud is defined and the penalties levied against violators, the overwhelming majority of employ-ees will likely be discouraged from becoming involved in such schemes. Nevertheless, some employees may not be so easily deterred. For this reason, educating employees about workers' compensation fraud should include informing employees of how they can be hurt by others within the company committing workers' compensation fraud. As mentioned in a previous section, workers' compensation fraud causes honest employees to shoulder an increased workload and may negatively affect wage increases of all employees by decreasing the company's overall profit margin. Employees should also be advised of how they can help curb fraud. Many workers' compensation insurers have established 24-hour anonymous hot lines for the reporting of actual or suspected workers' compensation fraud. If such a line is available, employers should educate employees of how the reporting system works and should place posters in visible locations indicating the telephone number as a constant reminder.

Another administrative control to combat workers' compensation fraud is the provision of monetary rewards or other incentives for the identification of persons who commit workers' compensation fraud. These programs generally allow the person reporting the fraudulent activity to remain anonymous among his or her coworkers. If administered by the company's workers' compensation insurer, the state workers' compensation board, or an independent entity, the identity of the whistle-blower can remain anonymous to the employer as well. If such monetary rewards are provided for identifying persons who commit workers' compensation fraud, that information should be posted so that all employees are aware of the reward program through a constant visible reminder.

10.7.7   Maintaining effective communication

Educating employees about workers' compensation fraud, the impact that it has on the company and individual employees, and the ways in which individual employees can help thwart workers' compensation fraud constitute effective communication. However, maintaining effective communication as a means of preventing workers' compensation fraud goes far beyond employee education efforts. To deter and detect workers' compensation fraud, the employer must effectively communicate with employees, medical care providers, and the company's workers' compensation insurance carrier. The following paragraphs describe the types of communication that can prove to be productive.

10.7.7.1   Employees

Just like any other type of relationship, effective communication, feelings of mutual respect, and good rapport between employers and employees help to prevent conflict and create a desire for both entities to please the other. An employee who views the relationship with his employer as favorable will be less likely to engage in acts, such as workers' compensation fraud, that may damage that relationship. To produce and sustain such a relationship, employers, managers, and supervisors must make a concerted effort to truly know employees on an individual basis and to be aware of their concerns and problems both on and off the job. It is only then that employers can build a mutual trust and respect with employees that can withstand the inevitable tribulations.

Although an ancillary result of maintaining effective communication with employees, employers become aware of activities such as sports or second jobs that employees may have outside of their primary work environment. This knowledge may prove beneficial if an employee submits a questionable workers' compensation claim and is known to be involved in contact sports or a physically demanding second job, as an employee may try to receive workers' compensation benefits for an injury that has occurred outside of work.

In addition to maintaining good relations in the normal course of business, employers should be cognizant of the fact that immediately following a work-related injury there is an opportunity for the employee/employer relationship to be strengthened or impaired based almost exclusively on how the employer reacts and communicates with the employee. Following an injury, the injured employee is likely to be worried and distressed, not knowing if his medical bills will be paid in full. If the injury results in time off work, the employee may be distraught about when he will see a paycheck and how much it will be. The injured employee may also be concerned that he may be replaced during time off work or that he will be viewed by his employer as a problem employee. If the employer effectively communicates with the injured employee and conveys answers to questions that the employee may be afraid to ask, the injured employee is likely to view the employer as an advocate on his behalf, easing the anxiety brought about by an uncertain future. Conversely, if the employer is quick to ascribe blame for an injury, is curt and abrupt, or simply does nothing more than complete the First Report of Injury Form and send the employee to the doctor, there is a much greater chance that the employee will view his relationship with his employer as uncertain, leading to anxiety about pay, employment status, and how he will be perceived by others. This anxiety is prone to increase if the employee is off work due to the injury and the employer does not maintain regular communication, as the employee may begin to feel abandoned.

This anxiety and feeling of being abandoned may create a desire for the injured employee to seek retribution or perceived self-preservation through trying to profit financially by abusing the workers' compensation system. After all, while the injured employee is off work, the television is bombarding him or her with commercials encouraging injured employees to question if the benefits paid are truly all to which they are entitled. Hence, effective and constant communication with employees is particularly important following an injury, especially while the injured employee is off work due to an injury.

Consequently, employers should stay in constant contact with employees who are off work due to a work-related injury, being empathetic and offering to do things that promote good relations between the employee and management. For example, if an employee is off work due to a work-related back injury, the employee's supervisor or a member of senior management could offer to pick up groceries and deliver them to the injured employee's home.

Although important, effective communication with employees following an injury is not limited to that which is intended to maintain good relations. It can also include making employees aware that the company has a strong awareness of the potential for workers' compensation fraud. Whereas questioning the legitimacy of a claim in a confrontational manner with a claimant would likely create an immediate adversarial relationship between the employer and the claimant, the company's awareness of the potential for fraud can be communicated in a subtle manner. One method of doing this is to encourage employees to report physicians who are charging for treatments, tests, or rehabilitation that the claimant believes are unnecessary. By communicating with the employee in this manner, the employer is not only communicating its rigid stance against workers' compensation fraud but is strengthening the relationship with the employee by placing trust in him or her to report suspected fraud relating to his or her own claim.

In addition to claimants, effective communication with the coworkers of claimants is important, as well. If a claim is questionable, employers should talk with the claimant's coworkers regarding statements that the claimant may have made prior to the injury. Often employees will mention injuries that are not work-related and later submit a workers' compensation claim indicating that the injury was incurred on the job. Furthermore, if a claimant is off work due to a work-related injury, the close-knit coworkers of the claimant may be asked to speak with the claimant to determine when he intends to return to work.

10.7.7.2 Medical care providers

It is also important to maintain effective communication with the treating physicians of the claimant. One very effective method is to accompany claimants to their initial and subsequent physician visits. If unable to accompany the claimant to physician visits, the employer should contact the treating physician regularly to ascertain the medical status of the claimant, the treatment plans, and the claimant's progress toward recovery, as well as personal observations of the claimant. In addition to advising the employer of medical-related information, such as the claimant's medical status, the anticipated date of the claimant's return to work, and clarification of any temporary physical restrictions, the physician may also be able to offer observations not derived from a medical diagnosis. These nonmedical observations may include if the employee had significant dirt or grease under his fingernails, which would indicate that he was performing other work, or if the employee was wearing a uniform or driving a vehicle, which would indicate that he was working for another company. The treating physician may also be able to inform the employer if the claimant has missed any scheduled treatments or examinations. Although it may seem inconceivable that a claimant who is committing workers' compensation fraud would be so brazen as to flaunt his activity, he may view the physician's office as a safe haven.

10.7.7.3 *Insurance carrier*

Fraud can also be deterred and detected through maintaining good communication with the workers' compensation insurance carrier.

Employers should view the relationship with their workers' compensation insurer as a team effort. Both the company and the insurance carrier have (or should have) the same goals relative to workers' compensation claims — providing rapid benefits to employees who are legitimately injured on the job, deterring workers' compensation fraud, and denying benefits to those who perpetrate workers' compensation fraud. As both entities have the same goals, it is beneficial for both to work in concert. To do this there must be effective communication that enables both the employer and its insurance carrier to know what the other is doing relative to each suspect claim. If there is a duplication of efforts — for example, both the employer and the insurance carrier have spoken to the claimant regarding the cause of the injury — the claimant's responses should be compared to determine if they are consistent.

As an element of maintaining effective communication with its workers' compensation carrier, employers should inform their insurer as soon as suspicions arise concerning the legitimacy of a claim, keeping in mind that evidence supporting the denial of claims based on possible fraud can result from witness statements and physical evidence at the scene of an alleged injury. Although some employers wait until proof is obtained, this often hinders the investigative process. If a claim is suspect at the time it is submitted, on the First Report of Injury Form the employer should indicate "Questionable Claim — Please Investigate" and should followup with a telephone call to the insurer to provide a description of the reasoning behind questioning the claim's legitimacy.

In addition to the timely identification of questionable claims, effective communication with the workers' compensation insurance carrier includes employers becoming familiar with the insurance carrier's efforts to deter and detect fraud. By knowing what steps the insurance carrier takes to deter and detect fraud, the employer is better able to know what information should be provided to make the process more efficient. Furthermore, this communication enables the employer to become better educated about workers' compensation fraud, as most workers' compensation carriers are more than willing to educate employers and recommend efforts that employers can make to assist in deterrence and detection.

10.7.8 *Responding appropriately to claims*

One of the most critical roles of the employer is the timely reporting of all injuries. As evidence that points to possible fraud can result from witness statements and physical evidence at the scene of an alleged injury, employers should report all injuries as soon as possible. Generally insurance carriers prefer employers to report all work-related injuries within 24 hours after the claimant has notified the employer of an injury. The chances of obtaining adequate witness statements, photographs, and other evidence to counter a fraudulent claim decrease with time.

In addition to reporting injuries in a timely manner, supporting documentation should be forwarded to the workers' compensation insurance carrier as soon as possible. Although not every work-related injury warrants the same degree of internal investigation, certainly all work-related injuries should involve some investigation to determine how the injury occurred. Documentation beneficial in refuting claims believed to be fraudulent include written witness statements or other written statements from coworkers, such as statements from employees who are aware that the claimant had a preexisting injury or is engaged in other employment, or other statements that cause the employer to question the validity of a claim. Additionally, sketches or photographs of the accident scene may prove to be beneficial. Furthermore, employers who have closed-circuit televisions for security purposes may have videotaped evidence that confirms or incontrovertibly contradicts the claimant's description of an accident.

Many people are familiar with the type of videotape surveillance that shows claimants playing football after claiming to be unable to walk or lifting 100-lb objects at home after claiming that it caused too much pain to lift a 15-lb object at work. However, most surveillance that is performed for the purpose of investigating suspect claims is performed or contracted by the employer's workers' compensation insurance carrier. It is important for employers to remember that videotaped surveillance is not an undertaking for a novice. If the company elects to have surveillance work performed, it should use reputable surveillance investigators. Because of laws relating to the rights of privacy and rules of evidence, employers should be aware that the use of novice investigators might result in the inadmissibility of the evidence obtained.

10.7.9 Support of legislative efforts

Despite all of the efforts undertaken by the employer and by workers' compensation insurance companies, many suspected cases of workers' compensation fraud eventually find their way to an administrative law process in which attorneys for both sides argue their case based on current legislation. As such, the success or failure of efforts to prevent workers' compensation fraud, to prosecute the perpetrators of fraud, and to recover incurred losses is ultimately based on the statutes that compose workers' compensation law. Therefore, it is incumbent on all employers, insurance carriers, trade organizations, and other concerned entities to support antifraud legislation.

10.8 Summary

Although each state defines workers' compensation fraud somewhat differently, it is generically defined as knowingly making false or misleading statements with the intent to profit financially. It is an offense that not only includes that which is perpetrated by claimants but also includes the actions of physicians, attorneys, and even employers and insurance carriers themselves.

The financial impact of workers' compensation is astounding and is estimated to be anywhere from $3.5 billion to $5 billion per year. As with any other business, the costs incurred by the providers of workers' compensation insurance is passed on to its customers. This means that fraud causes insurance rates to increase for employers, who in turn must increase the cost of their goods and services or reduce salaries to compensate for their increased costs.

The good news is that employers are not powerless to fight workers' compensation fraud. Years of experience fighting workers' compensation fraud have enabled the insurance industry to assemble warning signs, or "red flags," which are used as indicators to initiate further investigation of claims to determine their legitimacy. These warning signs relate to the injury, claimant, and medical treatment provided to a claimant. Additionally, if armed with knowledge, employers are able to detect and deter workers' compensation fraud through selecting the right workers' compensation carrier, selecting the right physicians, designating a workers' compensation claims coordinator, implementing effective administrative controls, exercising caution in hiring and firing, educating employees, and maintaining effective communication with employees, medical care providers, and their insurance carrier. Last, employers and other concerned entities can make a difference by supporting antifraud legislation.

chapter eleven

Common and costly injuries

Contents

11.1 Introduction

Some injuries present themselves more often than others. Unfortunately, these injuries are not select but occur in all types of industries and businesses, and, to further complicate the scenario, their medical treatment can be prolonged and relatively expensive.

This chapter focuses on workers' compensation claims that result from poor workplace ergonomics or slips, trips, and falls. These types of injuries can financially undermine an organization's workers' compensation insurance program. But it does not have to be this way. Practical injury avoidance techniques and proven claims management practices can contain the prevalence of these types of injuries and the associated costs.

11.2   Ergonomics

It might be said that the topic of ergonomics causes more pain than the injuries do! From reading the continuous point/counterpoint debate over OSHA's attempt to promulgate an ergonomics standard, it's obvious that the topic of ergonomics is an emotional and controversial issue.

The issue both excites and incites. Few can review the facts and fiction surrounding ergonomics without developing an opinion. Add to that the controversy of government regulation, and confrontation is the one true absolute that results. It has been a hard fight that has been waged through Democratic- and Republican-controlled Congresses, through the routine and relatively rapid turnover of OSHA leadership, and through the rise and fall of occupational injury rates directly attributed to ergonomic hazards. In fact, as of the writing of this text, the proposed standard that was published in November 1999 met the same fate as its 1995 predecessor.

An initial Draft Ergonomic Protection Standard was issued in June 1994, and a more detailed draft was issued in March 1995. Many regard this draft as the most polarizing by OSHA to date in the chronology of the ergonomics standard promulgation effort. As a result, such strong political pressure was applied and maintained that both the Clinton administration and OSHA put the proposal on the back burner, citing strong Congressional and business-interest opposition. Congressional opposition was so strong that in June 1995 OSHA was prohibited from expending any funds to issue a proposed or final ergonomics standard or guideline. Known as the "ergo rider," this political appropriation chicanery precluded any progress on the standard during that fiscal year. It expired in October 1996, and soon thereafter OSHA chose a new approach, a four-pronged strategy for ergonomics: educational and outreach activities, study and analysis, enforcement, and rulemaking. However, this rulemaking strategy was also derailed with "unfunding" provisions by Congress during the summer of 2000.

Regardless of the political situation or whether or not there is a regulatory document in effect, American workers are sustaining ergonomic injuries, and those injuries are expensive workers' compensation claims. Evidence of this fact is witnessed daily during the loss control consultative visits conducted by the authors.

11.2.1   Terms and definitions

The word *ergonomics* comes from Greek: *ergo* means work and *nomos* means law; therefore, "ergonomos" means "law of work." Ergonomics, however, means different things to different people. To an ergonomist, it means designing the workplace to ensure that employees are not being harmed by their job. To an engineer, it means maximizing employee productivity and reducing the chance of injury. To an occupational health professional, ergonomics is a plan to encourage employees to report early symptoms of cumulative trauma disorders and ensure that medical intervention or treatment is applied. To the supervisor, ergonomics is a way to improve the employee's job satisfaction. And finally, to the employee, ergonomics is an indication that management is concerned about safety and welfare. It's all a matter of perspective and impact.

Practically speaking, ergonomics is the science of fitting the job to the worker. When there is a mismatch between the physical requirements of the job and the physical capacity of the worker, musculoskeletal disorders (MSDs) can result. For example, workers who must repeat the same motion throughout their workday, do their work in an awkward position, use a great deal of force to perform their jobs, repeatedly lift heavy objects, or face a combination of these risk factors are most likely to develop MSDs.

11.2.2   MSD, RSI, RMI, CTD, or OOS?

Musculoskeletal disorders have been labeled with numerous titles, but the manifestations and the debilitating results are the same. MSD is also known as repetitive stress injury (RSI), repetitive motion injury (RMI), cumulative trauma disorder (CTD), and occupational overuse syndrome (OOS). Each is an injury or disorder of the muscles, nerves, tendons, ligaments, joints, cartilage, or spinal discs. MSDs do not include injuries resulting from slips, trips, falls, or similar accidents. Examples of MSDs include carpal tunnel syndrome, tendonitis, sciatica, herniated disc, and low back pain.

11.2.3   What a pain!

Just how serious a problem are work-related MSDs? They are the most prevalent, most expensive, and most preventable workplace injuries in the country. In addition, work-related MSDs have become a leading cause of lost-workday injuries and workers' compensation costs. According to OSHA:

- MSDs account for 34% of all lost-workday injuries and illnesses.
- More than 620,000 lost-workday MSDs are reported each year.
- MSDs account for $1 of every $3 spent for worker's compensation.
- Carpal tunnel syndrome (CTS), one form of MSD, results on average in more days away from work than any other workplace injury. The median number of days away from work for CTS is 25 days, as compared to 17 days for fractures and 20 days for amputations.
- Workers with cases of severe injury can face permanent disability that prevents them from returning to their jobs or handling simple, every-day tasks like picking up their child, combing their hair, or pushing a shopping cart.

Many direct, indirect, and hidden costs are associated with these MSD injuries, as shown in Figure 11.1. Don't forget that an injured worker's absence requires a new employee to be introduced into the workplace, and that means that the portion of the work force that proportionately sustains a higher number of injuries is increased. Also, realize that a decrease in employee morale contributes to increased employee turnover, which in turn increases the size of this high-risk portion of the work force, which can potentially increase the number of workers' compensation claims.

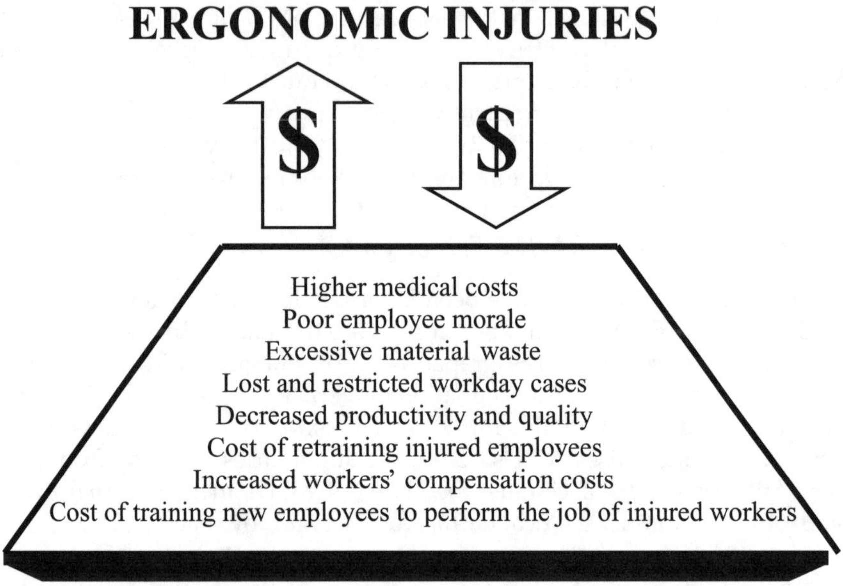

ERGONOMIC INJURIES

Higher medical costs
Poor employee morale
Excessive material waste
Lost and restricted workday cases
Decreased productivity and quality
Cost of retraining injured employees
Increased workers' compensation costs
Cost of training new employees to perform the job of injured workers

Figure 11.1

11.2.4   What causes work-related MSDs?

These occupational injuries occur where there is a mismatch between the physical requirements of the job and the physical capacity of the worker. Prolonged exposure to ergonomic risk factors, particularly in combination or at high levels, is likely to cause or contribute to any MSD or aggravate the severity of a pre-existing MSD. The longer and more frequent the exposure to ergonomic risk factors, the longer the time needed to recover from exposure to them. Risk factors include force, repetition, awkward postures, static postures, vibration, and cold temperatures.

Having just reviewed the list of risk factors, mentally inventory the jobs that exist in your company. Do any of them contain some — or even worse, all — of these factors? Have any of your employees complained of low back pain or numbness in their extremities?

The cause of musculoskeletal disorders can be elusive. Was the source of the injury at work or off work? Contributing or causative factors in the workplace are not so elusive, however, because they can usually be observed and quantified. An employee picks up 500 totes each shift, or an employee grasps, twists, and inserts the widget into the container at a rate of 100 times per minute. That same employee may also be exposed to several risk factors for the 16 hours that he or she is not in the workplace, and these risk factors are very difficult to quantify. Which is causing more damage and who is responsible?

Another consideration is the intricacies and multiple variables involved with each individual's injury. For example, did the employee use equipment improperly? Did the employee's general level of fitness contribute to the injury? Was the employee's diet deficient? Did lack of job satisfaction cause the injury? Was the level of stress at home sufficient to generate an injury?

The simple rule is that if it happens at work or because of work, then it probably is compensable. Thus, did the employee clock in that day with a low-pain threshold injury that became a high-pain threshold injury because of forceful, repetitive work-related motions, or was the injury an effect of the natural aging process? These are examples of situations that confront employers daily, and it's frustrating — to the employer, the employee, the insurance provider, and the medical community.

Ergonomic injuries are insidious for generating suspicion regarding the validity of the claim because quite often there are no physical manifestations of the injury. When you hurt your lower back it does not swell or bleed or appear distorted, but it definitely hurts! No one can feel or see your pain; they can only observe your behavior. If they have any reason to question your integrity, then the injury becomes the focus of their distrust.

Ergonomic injuries require longer recovery periods than most other injuries, and they tend to recur. If there ever was good reason to have a solid modified duty program, this is it. Allowing an employee to return to the same occupational environment (job) that still contains the unabated ergonomic risk factors, which probably contributed to the injury, is a virtual guarantee for recurrence of that injury.

11.2.5   What are possible solutions, if any?

MSDs are avoidable and are often very easy to prevent. Placing a book under a computer monitor, or padding a tool handle are typical of the fixes used in ergonomics programs. Companies that implement an ergonomics program achieve solutions that fit the work to the worker. Thousands of employers have adopted them. In fact, a study by the General Accounting Office found employers' ergonomics programs effective at reducing injuries. Practical experience in solving ergonomics problems is plentiful. If you walk through your workplace and closely observe, you will notice how employees have already adapted their workstation or tools to their bodies. Ergonomic interventions may involve simple solutions such as

- Adjusting the height of working surfaces to reduce long reaches and awkward postures
- Putting work supplies and equipment within comfortable reach
- Providing the right tool for the job and the right handle for the worker
- Varying tasks for workers (job rotation)

- Encouraging short, authorized rest breaks
- Reducing the weight and size of items workers must lift
- Providing mechanical lifting equipment so workers won't strain their backs lifting heavy items by themselves
- Using telephone headsets
- Providing ergonomic chairs or stools
- Supplying antifatigue floor mats

Controlling MSDs is only one of many reasons for establishing ergonomic programs. Well-run ergonomic programs have been effective in improving quality and productivity, improving employee satisfaction, and reducing turnover and absenteeism. They can also help improve the competitiveness of a company. Setting up an effective program can help control skyrocketing costs of injuries. The price of putting a program into action will be minor compared to the money saved on workers' compensations costs. Money will also be saved on medical bills because injuries will be identified, treated, and stopped before the case progresses to a more expensive stage.

It is recommended that ergonomic programs contain the following seven elements, which closely resemble those in OSHA's model safety and health program:

1. Looking for signs of a potential musculoskeletal problem in the workplace, such as frequent worker reports of aches and pains, or job tasks that require repetitive, forceful exertions
2. Showing management commitment in addressing possible problems and encouraging worker involvement in problem-solving activities
3. Offering training to expand management and worker ability to evaluate potential musculoskeletal problems
4. Gathering data to identify jobs or work conditions that are most problematic, using sources such as injury and illness logs, medical records, and job analyses
5. Identifying effective controls for tasks that pose a risk of musculoskeletal injury and evaluating these approaches once they have been instituted to see if they have reduced or eliminated the problem
6. Establishing healthcare management to emphasize the importance of early detection and treatment of musculoskeletal disorders for preventing impairment and disability
7. Minimizing risk factors for musculoskeletal disorders when planning new work processes and operations since it is less costly to build good design into the workplace than to redesign or retrofit later

Examples of typical applications of these elements that employers have taken to address MSDs are listed in Figure 11.2. Each of these is a commonsense evaluation and subsequently inexpensive and uncomplicated approach to eliminating or controlling the hazard.

- Look at injury and illness records to find jobs where problems have occurred
- Talk with workers to identify specific tasks that contribute to pain and lost workdays
- Ask workers what changes they think will make a difference
- Use employee comments to determine what improvements need to be made and then implement the employee recommendations
- Encourage workers to report MSD symptoms and establish a medical management system to detect problems early
- Find ways to reduce repeated motions, forceful hand exertions, prolonged bending, or working above shoulder height
- Reduce or eliminate vibration and sharp edges or handles that dig into the skin
- Rely on equipment, not backs, for heavy or repetitive lifting
- Use educational programs to train employees and management about ergonomic techniques to prevent and correct MSDs
- Provide ongoing evaluations to ensure that improvements are constantly made

Figure 11.2

There really is nothing mystical about ergonomics. The difficulty and controversy center on the origin of the injury and the ultimate liability. In the meantime, however, these injuries are occurring at an alarming rate and are driving the cost of workers' compensation ever higher. The bottom line of your business is being affected, but these injuries can be controlled.

11.2.6 Loss control consultative visit observations/recommendations

On a daily basis the authors visit insureds to assist in the identification and control of workplace hazards that may contribute to occupational injuries or illness. Thousands of businesses have been inspected. Unique samples of ergonomic hazards noted in those companies are described below. Included, also, are the recommendations provided to the employer. Review these situations and evaluate your workplace to assess whether similar hazards or methods of abatement are applicable.

Situation 1. A textile industry with approximately 20 employees is involved with the fabrication of leather and jersey utility gloves. The work force is predominantly female and their average age is late 40s to mid 50s. One employee is in her early 70s. The loyalty of these employees is unparalleled. Job rotation has been attempted, but to a limited degree. The current employers recently purchased the company and are not aware of any significant repetitive stress injuries, but they are very much aware of the occupational hazards in the industry and are open to all suggestions.

Recommendations — As discussed, the textile industry is notorious for producing ergonomic injuries. Add to this the fact that several of the employees in XYZ Glove Company are older workers and the conditions are prime for just such an occupational injury to occur. The control of this risk is a challenge and not so easily applied. Since engineering controls have been attempted, and to some extent have proven effective, the next level of mitigation is administrative controls. Administrative control strategies include (1) changes in job rules and procedures such as scheduling more rest breaks, (2) rotating workers through jobs that are physically tiring, and (3) training workers to recognize ergonomic risk factors and to learn techniques for reducing the stress and strain while performing their work tasks. Since these controls do not eliminate hazards, management must assure that the practices and policies are followed. Common examples of administrative control strategies for reducing the risk of work-related musculoskeletal disorders (WMSDs) include the following:

- Reducing shift length or curtailing the amount of overtime
- Rotating workers through several jobs with different physical demands to reduce the stress on limbs and body regions
- Scheduling more breaks to allow for rest and recovery
- Broadening or varying the job content to offset risk factors (repetitive motions, static and awkward positions, etc.)
- Adjusting the work pace to relieve repetitive motion risks and give the worker more control of the work process
- Training in the recognition of risk factors for WMSDs and instruction in work practices that can ease the task demands or burden

Considering the complexity of this challenge and the fact that ergonomic injuries are so elusive and expensive, it may be advisable to seek the assistance of a trained, professional ergonomist.

Situation 2. A medium-sized metal stamping industry has experienced a continuous incidence of shoulder and back strains. The requirement to move heavy objects has been alleviated through the use of forklifts, two-wheel handcarts, pallet jacks, and overhead cranes. The most common cause of these injuries is the "in-and-around" movement of materials in close proximity to the presses. Engineering fixes have been applied to the fullest extent, but injuries continue. This company has a medium to high employee turnover rate.

Recommendations — The best effort to reduce the incident rate of ergonomic injuries may lie in training/education. All too often our level of safety awareness becomes minimal over time, particularly in regard to common tasks. Many of the strains incurred by the employees were sustained during common tasks. Contact a local health provider (physical therapist, chiropractor, sports medicine trainer) and explore the opportunity for a class on lifting safety — a discussion that explains the mechanics of the lower spine and the impact of compressive forces during a lift. Maybe the healthcare provider will oblige without any fee in return for the publicity. Perhaps a local/regional safety consultant could be contracted to prepare and present a *site-specific* series of ergonomic awareness classes. If these prove unsuccessful, select the appropriate training material available through the Loss Control Department and present it in an employee safety meeting. The important point is to discuss lifting safety, soon and often.

The frequency rate of ergonomic injuries may be reduced by more selectively screening new hires during the application process. Pre-existing injuries may be sufficient evidence to preclude assigning an individual to a high-risk job without adequate protection. Protective measures range from redesigning the job to providing appropriate training; hence, the recommendation above is made.

Situation 3. A company involved in the fabrication of woven plastic bags is experiencing a steady increase in the incidence of strains to employees' upper extremities. Material handling of the bulk raw product is not the source of injuries. Rather, the remaining production processes (hand assembly, sewing, and bundling) involve a significant amount of manual involvement and repetitive motion exposure. The hand assembly operation involves inserting the plastic film sleeves into the woven fabric sleeves. According to the plant supervisor, employees perform this operation 1800 to 3000 times per day. The sewing operation involves folding one end of the hand-assembled bag and sewing that end closed using an industrial sewing machine. It is estimated that employees perform this operation 9,000–10,000 times per day. The company is locally owned and operated and competes for labor with regional and national corporations. Employees are paid per production; turnover is high.

Recommendations — Another factor that can contribute to the acceleration of ergonomic injuries (repetitive stress injuries) is piece-rate work. When an individual understands that the speed of production equals higher pay, then that employee is motivated to work faster. This can be detrimental if the production task involves repetitive motion, for that same employee may continue to work through pain in order to achieve production. The pain is a signal that the employee's soft tissue (muscle or joint connective tissue) has been strained. Rest is the first and most effective medical treatment, but if the employee ignores the pain and continues to flex the joint, then the physical injury can worsen to a significant extent. The result quite often manifests itself in occupational injuries such as carpal tunnel syndrome, tendonitis, or tenosynovitis. Granted, a business must remain operationally competitive to be profitable, but it appears that the piece-rate pay scale may be indirectly creating a drain on human resources and affecting the cost of workers' compensation insurance. Consider reverting to a fixed pay scale, at a locally competitive rate, but in conjunction with this change also hold employees to a higher standard of quality. The net result should be less occupational injury, lower employee turnover, higher morale, and even an equal or higher profit margin.

Encourage the consistent use of the overhead crane to lift rolls of woven fabric onto the printing machine. If this system is too cumbersome, encourage your creative employees to devise and then fabricate a mechanical lift method that is easier to use. In fact, institute a suggestion awards program for successful solutions to all the ergonomic challenges present in the production area.

Situation 4. A small company that cuts, heats, and forges rebar into industrial screws was visited early within their initial policy year. It was determined during this visit that a current employee had previously developed carpal tunnel syndrome as the result of one of the production tasks in the forging operation. That individual had subsequently undergone surgical repair and therapy and was back at work — doing the same process! Nothing had changed except that the employee's carpal tunnel had been surgically invaded for relief. The production process had not been changed.

Recommendations — The greatest risk of injury within the operation is most likely ergonomic, specifically repetitive motion injury (RMI). In fact, the very expensive injury incurred was a classic example of an RMI. Several options are available to avoid these types of injuries. The most desirable and usually most expensive is to simply engineer out the risk — modify the machine or workstation to eliminate the repetitive motion. To accomplish this it is recommended that an industrial engineer assess your operations. A less expensive, but also less effective, method is to rotate personnel through job tasks so that no one individual is exposed to prolonged periods of repetitive motion. As discussed, this may not be acceptable to the employees, particularly if they are paid by piecework. However, this is a manager's decision, and the welfare of the individual and the business must be considered. Finally, an even less effective but very inexpensive method of preventing RMIs is to educate all employees regarding the situation. If they have a joint or body part that is sustaining pain and injury from a repetitive motion, then they should know to tell their supervisor. In regard to the employee who sustained the RMI, it is strongly recommended that he or she never be allowed to continue working at the same workstation, performing the same repetitive tasks that induced his or her initial injury. The possibility for reinjury is great and the liability would approach gross negligence.

11.2.7 Summary

Setting aside all of the disputes involved in the ergonomics debate, what remains is a high incident rate of ergonomic injuries that are filed against workers' compensation insurance. The injuries are decimating the productivity of American businesses and negatively affecting their workers' compensation rates.

What can you do? What should you do? Obtain copies of informative literature from OSHA and NIOSH and become educated on ergonomics. Research data from other sources, those that may be diametrically opposed to ergonomic regulation. It doesn't matter whose literature you review — just learn about ergonomics so you can reduce the frequency and severity of these injuries in your company.

11.3 Slips, trips, and falls

"Who left that there? Where is my hammer? This place is a mess. We need to do some cleaning and get organized. I spend more time looking for my tools than I spend using my tools!" Sound familiar? Perhaps this situation is applicable to your business operation.

Back injuries or joint sprains and strains that result from slipping, tripping, or falling are not considered ergonomic injuries. But recovery from these soft tissue injuries is just as slow and expensive. Similar to ergonomic injuries, these can also be contained without tremendous effort. Simply put, clean house. The absence of occupational housekeeping contributes to numerous injuries that could have been prevented by simply putting an item where it belongs. Again, there's nothing mysterious about the causes or prevention of these types of injuries.

Few of us could argue that a neat and orderly facility is not an admirable goal, but there is much more to be gained by achieving this goal than just being neat and orderly — good occupational housekeeping will reduce your losses and increase your profits.

11.3.1 Loss control

"The mechanic *slipped* on the oily concrete surface. When the employee came around the corner he *tripped* over a box that someone had left on the floor. She *fell* down the stairs when she stepped on some books that were temporarily stored there." These slips, trips, and falls resulted from poor housekeeping, and these descriptions are direct quotes from 1st Reports of Injuries submitted by business owners.

We all realize that slipping, tripping, and falling too often result in painful injuries such as sprains, strains, fractures, bruises, and lacerations. Some can be permanently disabling or even fatal. How about puncture wounds or loss of teeth? Ever had an object stuck in your eye as the result of falling into a pile of "stuff"? Even worse, usually the accident victim will sustain several of these injuries during the slipping, tripping, or falling, such as fracturing both wrists and herniating a lumbar disk while attempting to break a fall from a ladder.

Although not too often a workers' compensation claim, but still a possibility, individuals can also receive burns from fires that originate in improperly stored flammable materials. These injuries can result either from attempting to extinguish the blaze with a portable fire extinguisher or during the evacuation of the structure.

The costs of these accidents and the subsequent loss incurred by your business is obvious.

11.3.2   Business enhancement

First impressions are critical. Have you ever visited a messy establishment? What was your opinion of the management, product, or quality of service? Your business's public image is largely determined by housekeeping.

The efficiency of internal operations dictates productivity. No one is efficient if they must halt operations to extinguish fires, clear passageways for the movement of forklifts, or properly store materials the second time because the first attempt was sloppy. Double work and remedial actions are unproductive activity.

Damaged product or equipment is a drain on company profits. How many flat tires have occurred in the yard because of debris left there? Does your company experience any clipping of storage racks because aisles are too constricted? Has a product ever missed shipment because it was not stored in its proper location? Does it seem that tools are constantly being replaced, only to have the originals turn up somewhere else on the premises? Good housekeeping can minimize these losses and increase business efficiency.

Employee morale is also heightened when a business is neat and orderly. The frustration of working in a mess is eliminated and the employees take pride in their surroundings. The benefits from this alone are immeasurable.

Granted, we're all very busy, so to assist in the improvement of your occupational housekeeping, the following considerations are offered.

11.3.3   Attaining and maintaining orderliness

Clean as you go. Everyone is responsible for housekeeping in his or her work area; it is a continuous process and a component of his or her job. End-of-shift cleanup chores are essential, but waste and residue should be addressed periodically through the shift; otherwise, the workplace may become a maze of trash piles. Include this task in the job description.

What the boss checks gets done. Incorporate the use of housekeeping checklists in the routine facility inspections that occur. Sample checklists are available free at OSHA (www.osha.gov) and SafetyInfo (www.safety-info.com) on the Internet.

Also, hold supervisors and employees responsible for housekeeping. Comments included in spot reports and performance appraisals validate the importance of such standards. Remember to give credit where it is due. Accentuate the positive and eliminate the negative.

You might also consider contracting custodial services or creating a position for such an employee. This is only a partial solution, however. Read the first recommendation again.

11.3.4 Regulatory requirements

OSHA standards address housekeeping, check 29 CFR 1910.22 General Requirements, partially quoted below.

> (a) Housekeeping. (1) All places of employment, passageways, storerooms, and service rooms shall be kept clean and orderly and in a sanitary condition.
>
> (a) (2) The floor of every workroom shall be maintained in a clean and, so far as possible, a dry condition. Where wet processes are used, drainage shall be maintained, and false floors, platforms, mats, or other dry standing places should be provided where practicable.
>
> (a) (3) To facilitate cleaning, every floor, working place, and passageway shall be kept free from protruding nails, splinters, holes, or loose boards.
>
> (b) Aisles and passageways. (1) Where mechanical handling equipment is used, sufficient safe clearances shall be allowed for aisles, at loading docks, through doorways and wherever turns or passage must be made. Aisles and passageways shall be kept clear and in good repairs, with no obstruction across or in aisles that could create a hazard.
>
> (b) (2) Permanent aisles and passageways shall be appropriately marked.

Another OSHA standard, 29 CFR 1910.38 Employee Emergency Plans and Fire Prevention Plans, also addresses housekeeping. The applicable citation is listed below:

> (b) (3) Housekeeping. The employer shall control accumulation of flammable and combustible waste materials and residues so that they do not contribute to a fire emergency.

11.3.5   Summary

Good occupational housekeeping can save you money, can increase your profit margin, and is the law. There's also sage advice concerning house-keeping that has transcended several generations of our ancestors:

> Everything should have a place and everything should
> be in its place.

> Cleanliness is next to godliness.

Maintaining an orderly facility is simple — assign responsibility and accountability in the execution of daily operations. Everyone should partic-ipate, from clerical to production to maintenance to accounting. Do not allow a "messy Marvin" to increase your workers' compensation premium.

Appendix A

Sample
Safety Program
for
"Your Company"

A Guide to Preventing
Work-Related Injuries
Through Effective Safety Management

Preface

How to use this manual

To the employer:

The purpose of this safety and health manual is to establish standards for an industry-specific safety and health program. This manual is intended to serve as the basis for an employer integrated safety and health management program.

The essential elements of this program include top management's commitment and involvement; the establishment and operation of safety committees; provisions for safety and health training; first aid procedures; accident investigations; record keeping of injuries; and workplace safety rules, policies, and procedures.

If this manual meets the needs of your establishment, you may choose to use it exactly as written. If you have previously established and are maintaining a safety program, you may choose to continue to use your program, provided that the essential elements covered in this safety program are also addressed in your program. Use of all or part of this manual does not relieve employers of their responsibility to comply with applicable local, state, or federal laws.

It is likely that state and federal OSHA standards require additional written safety programs, depending upon the type of work performed. Although not addressed in this sample safety program, the Loss Control Department of "your insurance company" is able to assist in the development of these written programs. Commonly required written safety programs include a Hazard Communication Program, a Forklift Training Program, a Respiratory Protection Program, an Exposure Control Plan, a Lockout/Tagout Program, and other written safety programs that address specific hazards.

It is intended that this manual be enhanced and continuously improved by the employer. Any section of this manual may be modified by the employer to accommodate actual operations and work practices, provided that the original intent of that section is not lost. For example, if a safety committee meets weekly or quarterly instead of monthly, then Section II of the manual should be amended to accommodate this practice. If there is a safety rule, policy, or procedure appropriate for the work or work environment which has not been included, then a new safety rule, policy, or procedure should be added to improve the manual. Likewise, if a specific rule in the Safety Rules, Policies, and Procedures section does not apply because the equipment or work operation described is not used, then that specific rule should be crossed out or deleted from the manual.

Table of contents

Preface

Section I — Management Commitment and Involvement

Policy statement

Section II — Safety Committee

Safety committee organization
Responsibilities
Meetings
Safety committtee minutes

Section III — Safety and Health Training

Safety and health orientation
Job-specific training
Periodic retraining of employees

Section IV — First Aid Procedures

Emergency phone numbers
Minor first aid treatment
Nonemergency medical treatment
Emergency medical treatment
First aid training
First aid instructions

Section V — Accident Investigation

Accident investigation procedures
Accident investigation report

Section VI — Recordkeeping Procedures

For assistance in developing or improving your company's safety program, contact the Loss Control Professionals at your insurance company.

Section I

Management commitment and involvement

Policy statement

The management of this organization is committed to providing employees with a safe and healthful workplace. It is the policy of this organization that employees report unsafe conditions and do not perform work tasks if the work is considered unsafe. Employees must report all accidents, injuries, and unsafe conditions to their supervisors immediately. No such report will result in retaliation, penalty, or other disincentive.

Employee recommendations to improve safety and health conditions will be given thorough consideration by our management team. Management will give top priority to and provide the financial resources for the correction of unsafe conditions. Similarly, management will take disciplinary action against an employee who willfully or repeatedly violates workplace safety rules. This action may include verbal or written reprimands and may ultimately result in termination of employment.

The primary responsibility for the coordination, implementation, and maintenance of our workplace safety program has been assigned to:

Name: _____

Title: _____ Telephone: _____

Senior management will be actively involved with employees in establishing and maintaining an effective safety program. Our safety program coordinator, other members of our management team, or I will participate with you or your department's employee representative in ongoing safety and health program activities, which include:

- Promoting safety committee participation;
- Providing safety and health education and training; and
- Reviewing and updating workplace safety rules.

This policy statement serves to express management's commitment to and involvement in providing our employees a safe and healthful workplace. This workplace safety program will be incorporated as the standard of practice for this organization. Compliance with the safety rules will be required of all employees as a condition of employment.

_____   _____
Signature of CEO/President   Date

Section II

Safety committee

Safety committee organization

A safety committee has been established as a management tool to recommend improvements to our workplace safety program and to identify corrective measures needed to eliminate or control recognized safety and health hazards. The safety committee employer representatives will not exceed the amount of employee representatives.

Responsibilities

The safety committee will be responsible for assisting management in communicating procedures for evaluating the effectiveness of control measures used to protect employees from safety and health hazards in the workplace.

The safety committee will be responsible for assisting management in reviewing and updating workplace safety rules based on accident investigation findings, any inspection findings, and employee reports of unsafe conditions or work practices; and accepting and addressing anonymous complaints and suggestions from employees.

The safety committee will be responsible for assisting management in updating the workplace safety program by evaluating employee injury and accident records, identifying trends and patterns, and formulating corrective measures to prevent recurrence.

The safety committee will be responsible for assisting management in evaluating employee accident and illness prevention programs, and promoting safety and health awareness and coworker participation through continuous improvements to the workplace safety program.

Safety committee members will participate in safety training and be responsible for assisting management in monitoring workplace safety education and training to ensure that it is in place, that it is effective, and that it is documented.

Management will provide written responses to safety committee written recommendations.

Meetings

Safety committee meetings are held monthly and more often if needed.

Each committee member will be compensated at his or her hourly wage when engaged in safety committee activities.

Management will post the minutes of each meeting in a conspicuous place and the minutes will be available to all employees.

All safety committee records will be maintained for not less than three calendar years.

Safety committee minutes

Date of Committee Meeting: _____   Time: _____

Minutes Prepared By: _____   Location: _____

Members in Attendance

Name

_____

_____

_____

_____

Previous Action Items: _____

_____

_____

Review of Accidents Since Previous Meeting: _____

_____

_____

_____

Recommendations for Prevention: _____

_____

_____

Recommendations from Anonymous Employees: _____

_____

_____

Suggestions From Employees: _____

_____

_____

Recommended Updates To Safety Program: _____

_____

_____

Recommendations from Accident Investigation Reports: _____

_____

_____

Safety Training Recommendations: _____

_____

_____

Comments: _____

_____

_____

_____

_____

_____

_____

_____

_____

_____

Section III

Safety and health training

Safety and health orientation

Workplace safety and health orientation begins on the first day of initial employment or job transfer. Each employee has access to a copy of this safety manual, through his or her supervisor for review and future reference, and will be given a personal copy of the safety rules, policies, and procedures pertaining to his or her job. Supervisors will ask questions of employees and answer employees' questions to ensure knowledge and understanding of safety rules, policies, and job-specific procedures described in our workplace safety program manual. All employees will be instructed by their supervisors that compliance with the safety rules described in the workplace safety manual is required.

Job-specific training

- Supervisors will initially train employees on how to perform assigned job tasks safely.
- Supervisors will carefully review with each employee the specific safety rules, policies, and procedures that are applicable and that are described in the workplace safety manual.
- Supervisors will give employees verbal instructions and specific directions on how to do the work safely.
- Supervisors will observe employees performing the work. If necessary, the supervisor will provide a demonstration using safe work practices, or remedial instruction to correct training deficiencies before an employee is permitted to do the work without supervision.
- All employees will receive safe operating instructions on seldom-used or new equipment before using the equipment.
- Supervisors will review safe work practices with employees before permitting the performance of new, nonroutine, or specialized procedures.

Periodic retraining of employees

All employees will be retrained periodically on safety rules, policies, and procedures, and when changes are made to the workplace safety manual.

Individual employees will be retrained after the occurrence of a work-related injury caused by an unsafe act or work practice, and when a supervisor observes employees displaying unsafe acts, practices, or behaviors.

Section IV

First aid procedures

Emergency phone numbers

Safety Coordinator (Name) _____ (Phone) _____

Ambulance _____ Medical Clinic_____

Fire Department _____ Police _____

Clinic Name/Address _____

Designated first aid providers

At this facility, the following employees have received appropriate training and have been designated as first aid providers:

_____ [ ] First Aid [ ] CPR [ ] Bloodborne Pathogens

_____ [ ] First Aid [ ] CPR [ ] Bloodborne Pathogens

_____ [ ] First Aid [ ] CPR [ ] Bloodborne Pathogens

Minor first aid treatment

First aid kits are kept in the front office and in the employee lounge. If you sustain an injury or are involved in an accident requiring minor first aid treatment:

- Inform your supervisor IMMEDIATELY.
- Administer first aid treatment to the injury or wound if you are trained to do so. If not, obtain the assistance from a trained and certified first aid provider.
- Complete an Accident Investigation Report and First Report of Injury Form (IA-1) with your supervisor
- If a first aid kit is used, indicate usage on the accident investigation report.
- Access to a first aid kit is not intended to be a substitute for medical attention. Seek medical attention if warranted.

Nonemergency medical treatment

If you sustain an injury requiring treatment other than first aid:

- Inform your supervisor IMMEDIATELY. For nonemergency work-related injuries requiring professional medical assistance, management must first authorize treatment.
- Proceed to an authorized medical facility. If a member of a managed care program, you will be advised by the Safety Coordinator of which medical facilities are acceptable. Your supervisor will assist with transportation, if necessary.
- Complete an Accident Investigation Report and First Report of Injury Form (IA-1) with your supervisor.

Emergency medical treatment

If you sustain a severe injury requiring emergency treatment:

- Inform your supervisor as soon as possible.
- Call for help and seek assistance from a coworker.
- Use the emergency telephone numbers and instructions posted next to the telephone in your work area to request assistance and transportation to the local hospital emergency room.
- Complete an Accident Investigation Report and First Report of Injury Form (IA-1) with your supervisor.

First aid training

Each employee will receive at least cursory training and instructions from his or her supervisor on our first aid procedures.

Only appropriately trained and certified employees should render first aid or CPR. The extent of first aid provided should be limited to that which the first aid provider is appropriately trained.

First aid instructions

In all cases requiring emergency medical treatment, immediately call or have a coworker call to request emergency medical assistance.

Wounds:

Minor: cuts, lacerations, abrasions, or punctures

- Wash the wound using soap and water; rinse it well.
- Cover the wound using clean dressing.

Major: large, deep, and bleeding

- Stop the bleeding by pressing directly on the wound, using a bandage or cloth.
- Keep pressure on the wound until medical help arrives.

Broken bones:

- Do not move the victim unless it is absolutely necessary.
- If the victim must be moved, "splint" the injured area. Use a board, cardboard, or rolled newspaper as a splint.

Burns:

Thermal (Heat)

- Rinse the burned area, without scrubbing it, and immerse it in cold water; do not use ice water.
- Blot dry the area and cover it using sterile gauze or a clean cloth.

Chemical

- Flush the exposed area with cool water immediately for 15 to 20 minutes.

Eye injury:

Small particles

- Do not rub your eyes.
- Use the corner of a soft clean cloth to draw particles out, or hold the eyelids open and flush the eyes continuously with water.

Large or stuck particles

- If a particle is stuck in the eye, do not attempt to remove it.
- Cover both eyes with bandage.

Chemical

- Immediately irrigate the eyes and under the eyelids with water for 30 minutes.

Neck and spine injury:

- If the victim appears to have injured his or her neck or spine, or is unable to move his or her arm or leg, do not attempt to move the victim unless it is absolutely necessary.

Heat exhaustion:

- Loosen the victim's tight clothing.
- Give the victim "sips" of cool water.
- Make the victim lie down in a cooler place with the feet raised.

Section V

Accident investigation

Accident investigation procedures

An accident investigation will be performed by the supervisor at the location where the accident occurred. The safety coordinator is responsible for seeing that the accident investigation reports are being filled out completely, and that the recommendations are being addressed. Supervisors will investigate all accidents, injuries, and occupational diseases using the following investigation procedures:

- Implement temporary control measures to prevent any further injuries to employees.
- Review the equipment, operations, and processes to gain an understanding of the accident situation.
- Identify and interview each witness and any other person who might provide clues to the accident's causes, obtaining a written statement from each witness.
- Investigate causal conditions and unsafe acts; make conclusions based on existing facts.
- Complete the accident investigation report.
- Provide recommendations for corrective actions.
- Indicate the need for additional or remedial safety training.

Accident investigation reports must be submitted to the safety coordinator within 24 hours of the accident.

Accident investigation report

REPORT # _____

COMPANY: _____ ADDRESS: _____

1. Name of injured: _____ S.S. #: _____

2. Sex [ ] M [ ] F Age: _____ Date of accident: _____

3. Time of accident: _____ a.m. _____ p.m. Day of accident: _____

4. Employee's job title: _____

5. Length of experience performing this type of work: ___ (years) ___ (months)

6. Address of location where the accident occurred: _____

7. Nature of Injury, Injury Type, and Part of the Body Affected: _____

    _____

8. Describe the accident and how it occurred: _____

    _____

    _____

    _____

    _____

    _____

    _____

9. Cause of the accident: _____

    _____

    _____

    _____

    _____

    _____

10. Was personal protective equipment required? [ ] yes [ ] no
    Was it provided? [ ] yes [ ] no
    Was it being used? [ ] yes [ ] no (If "no," explain.)

    _____

    _____

    Was it being used as trained by supervisor or designated trainer? [ ] yes [ ] no
    If "no," explain.

    _____

11. Witness (es):

    _____ (attach statement)

    _____ (attach statement)

    _____ (attach statement)

12. Safety training provided to the injured? [ ] yes [ ] no If "no", explain._____

    _____

13. Interim corrective actions taken to prevent recurrence: _____

    _____

14. Permanent corrective action recommended to prevent recurrence: _____

    _____

    _____

15. Date of report _____ Prepared by _____

Supervisor (Signature) _____ Date _____

16. Status and follow-up action taken by safety coordinator: _____

    _____

    _____

Safety Coordinator (Signature) _____ Date: _____

Instructions for completing the accident investigation report

An accident investigation is not designed to find fault or place blame but is an analysis of the accident to determine causes that can be controlled or eliminated.

(Items 1–6) Identification: This section is self-explanatory.

(Item 7) Nature of Injury: Describe the injury, e.g., strain, sprain, cut, burn, fracture. **Injury Type:** First aid — injury resulted in minor injury/treated on premises; Medical — injury treated off premises by physician; Lost time — injured missed more than one day of work; No Injury — no injury, near-miss type of incident. **Part of the Body:** Part of the body directly affected, e.g., foot, arm, hand, head.

(Item 8) Describe the accident: Describe the accident, including exactly what happened, and where and how it happened. Describe the equipment or materials involved.

(Item 9) Cause of the accident: Describe all conditions or acts which contributed to the accident (i.e., *unsafe conditions* — spills, grease on the floor, poor housekeeping, or other physical conditions; and/or *unsafe acts* — unsafe work practices such as failure to warn, failure to use required personal protective equipment).

(Item 10) Personal protective equipment: Self-explanatory.

(Item 11) Witness (es): List name(s), address(es), and phone number(s). Get a written statement from each witness and attach it to the Accident Investigation Report.

(Item 12) Safety training provided: Was any safety training provided to the injured related to the work activity being performed?

(Item 13) Interim corrective action: Measures taken by supervisor to prevent recurrence of incident; i.e., barricading accident area, posting warning signs, shutting down operations.

(Item 14): Self-explanatory.

(Item 15): Self-explanatory.

(Item 16) Follow-up: Once the investigation is complete, the safety coordinator shall review and follow up the investigation to ensure that corrective actions recommended by the safety committee and approved by the employer are taken, and control measures have been implemented.

Section VI

Recordkeeping procedures

The safety coordinator will control and maintain all employee accident and injury records. Records are maintained for a minimum of three (3) years and include:

- **Accident Investigation Reports** (make copies from the one contained in Section V of this Sample Safety Program)
- **Workers' Compensation First Report of Injury (Form IA-1) Form** (obtain copies from your workers' compensation insurance carrier)
- **Log & Summary of Occupational Injuries and Illnesses (OSHA 200 Log)** (a copy of this form can be obtained by contacting the Publications Department of federal OSHA, or your state OSHA, if applicable.)

Appendix B

OSHA Self-inspection Checklist

Safety and health program

Do you have an active safety and health program in operation that deals with general safety and health program elements as well as management of hazards specific to your worksite?

Is one person clearly responsible for the overall activities of the safety and health program?

Do you have a safety committee or group made up of management and labor representatives that meets regularly and reports in writing on its activities?

Are you keeping your employees advised of the successful effort and accomplishments you and/or your safety committee have made in assuring they will have a workplace that is safe and healthful?

Have you considered incentives for employees or work groups who have excelled in reducing workplace injuries/illnesses?

Personal protective equipment

Are employers assessing the workplace to determine if hazards that require the use of personal protective equipment (for example, head, eye, face, hand, or foot protection) are present or are likely to be present?

If hazards or the likelihood of hazards are found, are employers selecting and having affected employees use properly fitted personal protective equipment (PPE) suitable for protection from these hazards?

Has the employee been trained on PPE procedures, that is, what PPE is necessary for a job task, when they need it, and how to properly adjust it?

Are protective goggles or face shields provided and worn where there is any danger of flying particles or corrosive materials?

Are approved safety glasses required to be worn at all times in areas where there is a risk of eye injuries such as punctures, abrasions, contusions, or burns?

Are employees who need corrective lenses (glasses or contacts) required to wear only approved safety goggles or protective goggles, or use other medically approved precautionary procedures in working environments having harmful exposures?

Are protective gloves, aprons, shields, or other means provided and required where employees could be cut or where there is reasonably anticipated exposure to corrosive liquids, chemicals, blood, or other potentially infectious materials? See 29 CFR 1910.1030(b) for the definition of "other potentially infectious materials."

Are hard hats provided and worn where danger of falling objects exists?

Are hard hats inspected periodically for damage to the shell and suspension system?

Is appropriate foot protection required where there is the risk of foot injuries from hot, corrosive, or poisonous substances, falling objects, and crushing or penetrating actions?

Are approved respirators provided for regular or emergency use where needed?

Is all protective equipment maintained in a sanitary condition and ready for use?

Do you have eye wash facilities and a quick drench shower within the work area where employees are exposed to injurious corrosive materials?

Where special equipment is needed for electrical workers, is it available?

Where food or beverages are consumed on the premises, are they consumed in areas where there is no exposure to toxic material, blood, or other potentially infectious materials?

Is protection against the effects of occupational noise exposure provided when sound levels exceed those of the OSHA noise standard?

Are adequate work procedures and protective clothing and equipment provided and used when cleaning up spilled toxic or otherwise hazardous materials or liquids?

Are there appropriate procedures in place for disposing of or decontaminating personal protective equipment contaminated with, or reasonably anticipated to be contaminated with, blood or other potentially infectious materials?

Flammable and combustible materials

Are combustible scrap, debris, and waste materials (oily rags, etc.) stored in covered metal receptacles and removed from the worksite promptly?

Is proper storage practiced to minimize the risk of fire including spontaneous combustion?

Are approved containers and tanks used for the storage and handling of flammable and combustible liquids?

Are all connections on drums and combustible liquid piping, vapor, and liquid tight?

Are all flammable liquids kept in closed containers when not in use?

Are bulk drums of flammable liquids grounded and bonded to containers during dispensing?

Do storage rooms for flammable and combustible liquids have explosion-proof lights?

Do storage rooms for flammable and combustible liquids have mechanical or gravity ventilation?

Is liquefied petroleum gas stored, handled, and used in accordance with safe practices and standards?

Are "NO SMOKING" signs posted on liquefied petroleum gas tanks?

Are liquefied petroleum storage tanks guarded to prevent damage from vehicles?

Are all solvent wastes and flammable liquids kept in fire-resistant, covered containers until they are removed from the worksite?

Are fuel gas cylinders and oxygen cylinders separated by distance and fire-resistance barriers while in storage?

Are fire extinguishers selected and provided for the type of materials in areas where they are to be used (Class A — ordinary combustible material fires; Class B — flammable liquid, gas, or grease fires; Class C — energized electrical equipment fires)?

Are appropriate fire extinguishers mounted within 75 feet of outside areas containing flammable materials, and within 10 feet of any inside storage areas for such materials?

Are extinguishers free from obstruction or blockage?

Are all extinguishers serviced, maintained, and tagged at intervals not to exceed 1 year?

Are all extinguishers fully charged and in their designated places?

Where sprinkler systems are permanently installed, are the nozzle heads so directed or arranged that water will not be sprayed into operating electrical switch boards and equipment?

Are "NO SMOKING" signs posted where appropriate in areas where flammable or combustible materials are used or stored?

Are safety cans used for dispensing flammable or combustible liquids at a point of use?

Are all spills of flammable or combustible liquids cleaned up promptly?

Are storage tanks adequately vented to prevent the development of excessive vacuum or pressure as a result of filling, emptying, or atmosphere temperature changes?

Are storage tanks equipped with emergency venting that will relieve excessive internal pressure caused by fire exposure?

Are "NO SMOKING" rules enforced in areas involving storage and use of hazardous materials?

Are all tools and equipment (both company and employee-owned) used by employees at their workplace in good condition?

Are hand tools such as chisels and punches, which develop mushroomed heads during use, reconditioned or replaced as necessary?

Are broken or fractured handles on hammers, axes, and similar equipment replaced promptly?

Are worn or bent wrenches replaced regularly?

Are appropriate handles used on files and similar tools?

Are employees made aware of the hazards caused by faulty or improperly used hand tools?

Are appropriate safety glasses, face shields, etc. worn while using hand tools or equipment that might produce flying materials or be subject to breakage?

Are jacks checked periodically to ensure they are in good operating condition?

Are tool handles wedged tightly in the head of all tools?

Are tool cutting edges kept sharp so the tool will move smoothly without binding or skipping?

Are tools stored in dry, secure locations where they won't be tampered with?

Is eye and face protection used when driving hardened or tempered spuds or nails?

Portable (power-operated) tools and equipment

Are grinders, saws, and similar equipment provided with appropriate safety guards?

Are power tools used with the correct shield, guard, or attachment recommended by the manufacturer?

Are portable circular saws equipped with guards above and below the base shoe? Are circular saw guards checked to assure they are not wedged up, thus leaving the lower portion of the blade unguarded?

Are rotating or moving parts of equipment guarded to prevent physical contact?

Are all cord-connected, electrically operated tools and equipment effectively grounded or of the approved double-insulated type?

Are effective guards in place over belts, pulleys, chains, and sprockets on equipment such as concrete mixers and air compressors?

Are portable fans provided with full guards or screens having openings 1/2 inch or less?

Is hoisting equipment available and used for lifting heavy objects, and are hoist ratings and characteristics appropriate for the task?

Are pneumatic and hydraulic hoses on power-operated tools checked regularly for deterioration or damage?

Lockout/tagout procedures

Is all machinery or equipment capable of movement required to be de-energized or disengaged and locked out during cleaning, servicing, adjusting, or setting up operations, whenever required?

Where the power disconnecting means for equipment does not also disconnect the electrical control circuit, are the appropriate electrical enclosures identified and is means provided to assure the control circuit can also be disconnected and locked out?

Is the locking out of control circuits in lieu of locking out main power disconnects prohibited?

Are all equipment control valve handles provided with a means for locking out?

Does the lockout procedure require that stored energy (mechanical, hydraulic, air, etc.) be released or blocked before equipment is locked out for repairs?

Are appropriate employees provided with individually keyed personal safety locks?

Are employees required to keep personal control of their key(s) while they have safety locks in use?

Is it required that only the employee exposed to the hazard place or remove the safety lock?

Is it required that employees check the safety of the lockout by attempting a startup after making sure no one is exposed?

Are employees instructed to always push the control circuit stop button immediately after checking the safety of the lockout?

Is there a means provided to identify any or all employees who are working on locked out equipment by their locks or accompanying tags?

Are a sufficient number of accident preventive signs or tags and safety padlocks provided for any reasonably foreseeable repair emergency?

When machine operations, configuration, or size requires the operator to leave his or her control station to install tools or perform other operations, and that part of the machine could move if accidentally activated, is such element required to be separately locked or blocked out?

In the event that equipment or lines cannot be shut down, locked out and tagged, is a safe job procedure established and rigidly followed?

Confined spaces

Are confined spaces thoroughly emptied of any corrosive or hazardous substances, such as acids or caustics, before entry?

Are all lines to a confined space containing inert, toxic, flammable, or corrosive materials valved off and blanked or disconnected and separated before entry?

Are all impellers, agitators, or other moving parts and equipment inside confined spaces locked out if they present a hazard?

Is either natural or mechanical ventilation provided prior to confined space entry?

Are appropriate atmospheric tests performed to check for oxygen deficiency, toxic substances, and explosive concentrations in the confined space before entry?

Is adequate illumination provided for the work to be performed in the confined space?

Is the atmosphere inside the confined space frequently tested or continuously monitored during conduct of work?

Is there an assigned safety standby employee outside the confined space, when required, whose sole responsibility is to watch the work in progress, sound an alarm if necessary, and render assistance?

Is the standby employee appropriately trained and equipped to handle an emergency?

Is the standby employee or other employees prohibited from entering the confined space without lifelines and respiratory equipment if there is any question as to the cause of an emergency?

Is approved respiratory equipment required if the atmosphere inside the confined space cannot be made acceptable?

Is all portable electrical equipment used inside confined spaces either grounded and insulated, or equipped with ground fault protection?

Before gas welding or burning is started in a confined space, are hoses checked for leaks, compressed gas bottles forbidden inside of the confined space, torches lighted only outside the confined areas, and the confined area tested for an explosive atmosphere each time before a lighted torch is to be taken into the confined space?

If employees will be using oxygen-consuming equipment such as salamanders, torches, and furnaces in a confined space, is sufficient air provided to assure combustion without reducing the oxygen concentration of the atmosphere below 19.5% by volume?

Whenever combustion-type equipment is used in a confined space, are provisions made to ensure the exhaust gases are vented outside the enclosure?

Is each confined space checked for decaying vegetation or animal matter which may produce methane?

Is the confined space checked for possible industrial waste that could contain toxic properties?

If the confined space is below the ground and near areas where motor vehicles will be operating, is it possible for vehicle exhaust or carbon monoxide to enter the space?

Electrical

Do you specify compliance with OSHA for all contract electrical work?

Are all employees required to report as soon as practicable any obvious hazard to life or property observed in connection with electrical equipment or lines?

Are employees instructed to make preliminary inspections and/or appropriate tests to determine what conditions exist before starting work on electrical equipment or lines?

When electrical equipment or lines are to be serviced, maintained, or adjusted, are necessary switches opened, locked out, and tagged whenever possible?

Are portable electrical tools and equipment grounded or of the double-insulated type?

Are electrical appliances such as vacuum cleaners, polishers, and vending machines grounded?

Do extension cords being used have a grounding conductor?

Are multiple plug adapters prohibited?

Are ground-fault circuit interrupters installed on each temporary 15- or 20-amp, 120-volt AC circuit at locations where construction, demolition, modifications, or alterations are being performed?

Are all temporary circuits protected by suitable disconnecting switches or plug connectors at the junction with permanent wiring?

Do you have electrical installations in hazardous dust or vapor areas? If so, do they meet the National Electrical Code (NEC) for hazardous locations?

Is exposed wiring and cords with frayed or deteriorated insulation repaired or replaced promptly?

Are flexible cords and cables free of splices or taps?

Are clamps or other securing means provided on flexible cords or cables at plugs, receptacles, tools, equipment, etc., and is the cord jacket securely held in place?

Are all cord, cable and raceway connections intact and secure?

In wet or damp locations, are electrical tools and equipment appropriate for the use or location or otherwise protected?

Is the location of electrical power lines and cables (overhead, underground, under floor, other side of walls) determined before digging, drilling, or similar work is begun?

Are metal measuring tapes, ropes, hand lines, or similar devices with metallic thread woven into the fabric prohibited where they could come in contact with energized parts of equipment or circuit conductors?

Is the use of metal ladders prohibited in areas where the ladder or the person using the ladder could come into contact with energized parts of equipment, fixtures, or circuit conductors?

Are all disconnecting switches and circuit breakers labeled to indicate their use or equipment served?

Are disconnecting means always opened before fuses are replaced?

Do all interior wiring systems include provisions for grounding metal parts of electrical raceways, equipment, and enclosures?

Are all electrical raceways and enclosures securely fastened in place?

Are all energized parts of electrical circuits and equipment guarded against accidental contact by approved cabinets or enclosures?

Is sufficient access and working space provided and maintained about all electrical equipment to permit ready and safe operations and maintenance?

Are all unused openings (including conduit knockouts) in electrical enclosures and fittings closed with appropriate covers, plugs, or plates?

Are electrical enclosures such as switches, receptacles, and junction boxes provided with tight-fitting covers and plates?

Are disconnecting switches for electrical motors in excess of two horsepower capable of opening the circuit when the motor is in a stalled condition, without exploding? (Switches must be horsepower-rated equal to or in excess of the motor hp rating.) Is low-voltage protection provided in the control device of motors driving machines or equipment that could cause probable injury from inadvertent starting?

Is each motor-disconnecting switch or circuit breaker located within sight of the motor control device?

Is each motor located within sight of its controller or the controller disconnecting means capable of being locked in the open position or is a separate disconnecting means installed in the circuit within sight of the motor?

Is the controller for each motor in excess of two horsepower rated in horsepower equal to or in excess of the rating of the motor it serves?

Are employees who regularly work on or around energized electrical equipment or lines instructed in the cardiopulmonary resuscitation (CPR) methods?

Are employees prohibited from working alone on energized lines or equipment over 600 volts?

Walking-working surfaces

General Work Environment

Is a documented, functioning housekeeping program in place?

Are all worksites clean, sanitary, and orderly?

Are work surfaces kept dry or are appropriate means taken to assure the surfaces are slip-resistant?

Are all spilled hazardous materials or liquids, including blood and other potentially infectious materials, cleaned up immediately and according to proper procedures?

Are combustible scrap, debris, and waste stored safely and removed from the worksite properly?

Is all regulated waste, as defined in the OSHA bloodborne pathogens standard (1910.1030), discarded according to federal, state, and local regulations?

Are accumulations of combustible dust routinely removed from elevated surfaces including the overhead structure of buildings, etc.?

Is combustible dust cleaned up with a vacuum system to prevent the dust from going into suspension?

Is metallic or conductive dust prevented from entering or accumulating on or around electrical enclosures or equipment?

Are covered metal waste cans used for oily and paint-soaked waste?

Walkways

Are aisles and passageways kept clear?

Are aisles and walkways marked as appropriate?

Are wet surfaces covered with nonslip materials?

Are holes in the floor, sidewalk, or other walking surface repaired properly, covered, or otherwise made safe?

Is there safe clearance for walking in aisles where motorized or mechanical handling equipment is operating?

Are materials or equipment stored in such a way that sharp projectives will not interfere with the walkway?

Are spilled materials cleaned up immediately?

Are changes of direction or elevation readily identifiable?

Are aisles or walkways that pass near moving or operating machinery, welding operations, or similar operations arranged so employees will not be subjected to potential hazards?

Is adequate headroom provided for the entire length of any aisle or walkway?

Are standard guardrails provided wherever aisle or walkway surfaces are elevated more than 30 inches above any adjacent floor or the ground?

Are bridges provided over conveyors and similar hazards?

Floor and Wall Openings

Are floor openings guarded by a cover, a guardrail, or equivalent on all sides (except at entrance to stairways or ladders)?

Are toe boards installed around the edges of permanent floor openings (where persons may pass below the opening)?

Are skylight screens of such construction and mounting that they will withstand a load of at least 200 pounds?

Is the glass in the windows, doors, glass walls, etc., which are subject to human impact, of sufficient thickness and type for the condition of use?

Are grates or similar-type covers over floor openings, such as floor drains, of such design that foot traffic or rolling equipment will not be affected by the grate space?

Are unused portions of service pits and pits not actually in use either covered or protected by guardrails or equivalent?

Are manhole covers, trench covers, and similar covers, plus their supports, designed to carry a truck rear-axle load of at least 20,000 pounds when located in roadways and subject to vehicle traffic?

Are floor or wall openings in fire-resistive construction provided with doors or covers compatible with the fire rating of the structure and provided with a self-closing feature when appropriate?

Stairs and Stairways

Are standard stair rails or handrails on all stairways having four or more risers?

Are all stairways at least 22 inches wide?

Do stairs have landing platforms not less than 30 inches in the direction of travel and that extend 22 inches in width at every 12 feet or less of vertical rise?

Do stairs angle no more than 50 and no less than 30 degrees?

Are step risers on stairs uniform from top to bottom?

Are steps on stairs and stairways designed or provided with a surface that renders them slip-resistant?

Are stairway handrails located between 30 and 34 inches above the leading edge of stair treads?

Do stairway handrails have at least 3 inches of clearance between the handrails and the wall or surface they are mounted on?

Where doors or gates open directly on a stairway, is there a platform provided so the swing of the door does not reduce the width of the platform to less than 21 inches?

Where stairs or stairways exit directly into any area where vehicles may be operated, are adequate barriers and warnings provided to prevent employees from stepping into the path of traffic?

Do stairway landings have a dimension measured in the direction of travel, at least equal to the width of the stairway?

Elevated Surfaces

Are signs posted, when appropriate, showing the elevated surface load capacity?

Are surfaces elevated more than 30 inches above the floor or ground provided with standard guardrails?

Are all elevated surfaces (beneath which people or machinery could be exposed to falling objects) provided with standard 4-inch toe boards?

Is a permanent means of access and egress provided to elevated storage and work surfaces?

Is required headroom provided where necessary?

Is material on elevated surfaces piled, stacked, or racked in a manner to prevent it from tipping, falling, collapsing, rolling, or spreading?

Are dock boards or bridge plates used when transferring materials between docks and trucks or rail cars?

Hazard communication

Is there a list of hazardous substances used in your workplace?

Is there a written hazard communication program dealing with Material Safety Data Sheets (MSDS), labeling, and employee training?

Is each container for a hazardous substance (i.e., vats, bottles, storage tanks, etc.) labeled with product identity and a hazard warning (communication of the specific health hazards and physical hazards)?

Is there a Material Safety Data Sheet readily available for each hazardous substance used?

Is there an employee training program for hazardous substances?

Does the training include:

- An explanation of what an MSDS is and how to use and obtain one?

- MSDS contents for each hazardous substance or class of substance?

- Explanation of "Right to Know"?

- Identification of where an employee can see the employer's written hazard communication program and where hazardous substances are present in their work areas?

- The physical and health hazards of substances in the work area, and specific protective measures to be used?

- Details of the hazard communication program, including how to use the labeling system and MSDSs?

Are employees trained in the following?:

- How to recognize tasks that might result in occupational exposure?

- How to use work practice and engineering controls and personal protective equipment and to know their limitations?

- How to obtain information on the types, selection, proper use, location, removal handling, decontamination, and disposal of personal protective equipment?

- Who to contact and what to do in an emergency?

Appendix C

Company Safety and Workers' Compensation Cost Control Profile

XYZ Company is dedicated to the prevention of employee injuries and the control of costs associated with workers' compensation. To this end, XYZ Company has developed and implemented numerous safety programs and management controls which are outlined in this profile. It is the intent of this profile to provide our prospective workers' compensation carrier with an accurate depiction of the degree to which employee safety is valued at XYZ Company.

Mr. David Smith (Human Resource Manager) has been assigned overall responsibility for the administration and management of company safety efforts and the control of workers' compensation claims costs. Mr. Smith is very well-informed and knowledgeable of safety-related regulatory requirements that relate to the work performed by XYZ Company and continuously strives for 100% compliance. Although tremendous strides have been made to increase the safety of employees at XYZ Company, Mr. Smith is eager to receive and implement recommendations offered by your company's loss control representatives.

In addition to Mr. Smith, XYZ Company has assigned a Department Safety Officer from each manufacturing department. These employees are not supervisory personnel but have been given the authority to correct other employees with respect to safety-related issues.

Through a strong emphasis on employee safety, XYZ Company had achieved a 0.74 experience modification factor in 1995. However, since then the company has doubled in size, moved to a new location and regrettably misplaced our focus. For an approximate two-year period, occupational safety was addressed less aggressively. The result was a loss history for the 1997/98 and 1998/99 policy years that the owners and management of XYZ

Company deemed to be unacceptable, yielding a current experience modification factor of 1.15. XYZ Company accepted that as a wake-up call and within the past year has made a concerted effort to promote safety, conduct safety training, and create a corporate culture in which safety is the top priority. An example of XYZ Company's commitment to safety is that there is no budget for safety. Instead, the prevailing philosophy is to purchase whatever safety-related equipment is needed, regardless of cost.

The formal (written) safety programs maintained by this company include a Lockout/Tagout Program, an Emergency Action Plan, Internal Fire Safety Inspections, a Hearing Conservation Program, a documented PPE Hazard Assessment, a Fire Prevention Plan, a Respiratory Protection Program, a Hazard Communication Program, and a Bloodborne Pathogens Program. Documentation maintained by XYZ Company relative to each safety program is maintained in a separate three-ring binder. Furthermore, each safety program has been written in a manner as to be specific to the company facility and the operations performed by employees. A copy of each of these safety programs will be provided upon request.

XYZ Company places a significant emphasis upon ongoing safety training. Most of the formal safety training is conducted by the state OSHA Division of Education and Training, as each year Mr. Smith arranges for representatives to come to XYZ every three weeks to conduct some form of safety training. This training generally lasts at least two hours for each topic. Whereas the training conducted by OSHA representatives is not site-specific, Mr. Smith follows each training session by an explanation of how the information applies to the specific work performed by employees.

In addition to the above training, the Shop Foreman is charged with the responsibility of conducting employee safety meetings. These meetings are conducted at an average frequency of weekly and generally use topics from an occupational safety periodical to which the company subscribes.

In addition to obtaining services of the state's OSHA Division of Education and Training for the aforementioned employee training, Mr. Smith routinely solicits mock OSHA compliance inspections from the Division of Education and Training and corrects alleged violations immediately.

Safety committee meetings are conducted at an average frequency of monthly and include Mr. Smith, the Department Safety Officers, and several guests. The guests are employees of the company who do not serve an official safety role and may include employees who do not appear to share the "safety-first" philosophy of the company, and they are invited to evidence the seriousness with which this company addresses safety.

The company has a safety incentive program that includes recognition awards of $10 – 20 per month to all employees in the absence of lost-time work injuries. Additionally, the safety committee has established a goal relative to work-related injuries for the year 2000 (no more than four lost-time injuries). If the company achieves that goal, each employee will receive an additional bonus, based upon the amount of time he or she has been employed by XYZ Company.

In addition to the safety incentive programs mentioned above, this company has a safety suggestion program. This program provides $10 to each employee who makes a suggestion that could improve occupational safety at XYZ Company.

Mr. Smith conducts a formal (documented) safety inspection of the facility at a monthly frequency. A copy of the inspection form will be provided upon request.

The company has a drug-testing program, which includes preemployment, random, and post-accident drug testing. Any employee who tests positive for drugs in conjunction with a preemployment test is not hired. Anyone who tests positive for drugs in conjunction with post-accident testing is immediately terminated. Anyone who tests positive for drugs in conjunction with random testing is required to attend a substance abuse class. If the same person tests positive for drugs subsequently, he/she is terminated.

The company has formal (written) safety rules and a structured disciplinary policy. The disciplinary policy includes verbal correction for the first violation and a written warning for the second violation. Termination of empolyment is possible for the third violation or for any serious safety policy violation. Employees are required to sign a copy of the safety rules.

The company has an accident investigation program that seeks to identify the root cause of each incident. Mr. Smith requires a completed First Report of Injury and a completed Accident Investigation Form to be on his desk within 24 hours of an injury-producing incident, and he forwards a copy of this form to each manager and company safety officer immediately.

Mr. Smith is currently in the process of developing a more detailed policy manual which will include sections relating to CPR Certifications, General Safety and Accident Prevention Topics, Workers' Compensation, Care of Equipment, Health Safety Protection, Hazardous Chemicals, Drug-Free Workplace, Weapons-Free Environment, Workplace Violence, and General Rules of Conduct. A copy of this policy manual will be provided to the Loss Control Department of XYZ Company's workers' compensation carrier for recommended revisions.

The company publishes a monthly employee newsletter, which always includes an article relative to employee safety. Copies of this newsletter will be provided upon request.

XYZ Company recognizes the benefits of using modified duty as a means of managing the cost of workers' compensation claims; the company promotes early return-to-work following a work-related injury by providing modified duty tasks that are consistent with the physician-imposed, temporary physical restrictions.

The minimum age for employment with XYZ Company is 18 years of age. Successful applicants are required to be high school graduates and have at least one year of prior work history in a similar industry. However, the work history requirement is waived if the applicant has graduated from a vocational school, having received training which is closely related to the work for which he/she is submitting an employment application.

All employees are required to submit to a preemployment drug test.

All employees are required to submit to a preemployment physical examination. In addition to traditional general physical examination items, the required physical includes a baseline hearing-level examination and an examination of lumbar strength/mobility.

Although one Shop Foreman oversees all 15 of the production/assembly workers in the shop, he has two subordinate supervisors who oversee these production and assembly operations in his absence.

XYZ Company has continued to have a low rate of employee turnover over the past 12 years. During this period, the employee turnover rate, including employees hired to facilitate expansion, has never exceeded 20%.

Personal protective equipment provided for employee use includes safety glasses (required at all times while in the shop area), hearing protectors (required to be worn in designated area), steel-toed safety shoes (required to be worn at all times while in the shop area), welding hoods, welding gloves, leather aprons (required to be worn when welding), and air-purifying respirators (required to be worn when painting).

A press brake is used in this operation and is protected by a light curtain that causes the machine to stop if the worker's hand crosses the light beam.

Welding curtains are provided to protect adjacent workers from eye injury hazards associated with welding.

Material-handling aids provided for employee use include three fork-lifts, four 2-ton jib cranes, pallet jacks, two-wheeled hand-trucks, and carts.

There are currently four employees who have received training and maintain certification to render basic first aid and CPR.

Several first aid kits are maintained onsite and include personal protective equipment for the control of exposure to bloodborne pathogens.

Two eyewash stations are provided in the shop area. These eyewash stations are located at opposite ends of the shop area and provide a continuous uninterrupted flow of water.

Appendix D

Sample Modified Duty Program

A. Purpose

When an employee is injured in the course of employment and is eligible for workers' compensation indemnity (wage replacement) benefits, the employee often does not return to work promptly due to poor communication between the company, the workers' compensation insurance provider, and the employee's treating physician. This leads to unnecessary costs for the company, as well as potential financial hardship for the injured employee. Therefore, for the benefit of this company and its employees, this Modified Duty Program has been implemented.

This Modified Duty Program is intended to assist in the return to meaningful work within the abilities of injured employees. Simultaneously, this Modified Duty Program is designed to prevent malingering and system abuses which may result from inactivity during periods away from work and to provide for the monitoring of injured employees to ensure that physician-imposed physical restrictions are followed. Furthermore, this Modified Duty Program has the objective of reducing the direct and indirect costs associated with work-related injuries and of reducing the cost of future workers' compensation premiums while maintaining consistent and acceptable levels of productivity.

B. Scope

This Modified Duty Program applies equally to all employees of this company, insofar as the injury sustained by the employee has been determined by the company's current workers' compensation insurance carrier to be work-related and compensable under current workers' compensation legislation and insofar as work-related tasks which are within the physical limitations of the employee are available with reasonable accommodations made by the company, the injured employee, or both. Additionally, it is notable that the applicability of modified duty assignments is dependent upon the physical restrictions imposed by the treating physician, being limited to a specified, temporary period which does not exceed the maximum allowable period established by this Modified Duty Program.

C. Management commitment

When an employee of this company sustains a work-related injury or illness which is compensable under current workers' compensation legislation, and the treating physician releases the employee to return to work with temporary physical restrictions which preclude the injured employee from performing his or her regular job duties, without modifications, this company will make all reasonable efforts to enable the injured employee to return to work within the temporary, physician-imposed physical restrictions. To this end, this company has established written procedures and has assigned a management representative to serve as program coordinator (workers' compensation claims coordinator). Additionally, through the implementation of this Modified Duty Program, this company will conduct periodic employee training relative to modified duty, will maintain relevant documentation, and will conduct periodic evaluations of the program to determine its effectiveness. Furthermore, this company is committed to ensuring the consistent and equitable application of this Modified Duty Program.

To ensure appropriate and consistent application of this Modified Duty Program, this company's designated workers' compensation claims coordinator is responsible for the following:

- Reporting all potential workers' compensation insurance claims to the workers' compensation insurance carrier immediately.
- Maintaining documentation relative to the Modified Duty Program (to include training records, functional job descriptions, and records of potential and previously implemented modified duty assignments).
- Conducting training of employees relative to the company's Modified Duty Program.
- Maintaining effective communication with supervisors and managers to ensure that their designated responsibilities are being fulfilled.
- Maintaining effective communication with workers' compensation insurance claims representatives regarding claims to ensure knowledge of claim status and to advise of the injured employee's work/employment status, etc.
- Maintaining effective communication with employees' treating physicians regarding lost-time injuries and injuries involving temporary physical restrictions, as to obtain written temporary, physician-imposed physical restrictions for employees who have sustained a work-related injury.

- Coordinating the timely implementation of modified duty assignments following work-related injuries for which modified duty is applicable.
- Maintaining a working list of potential modified duty assignments that are available for employees who are released to return to work on a modified duty status.
- Adapting modified duty assignments as the injured employee's medical condition improves and physician-imposed physical restrictions are changed.
- Returning the injured employee to his/her regular job assignment when the employee is released by the treating physician to return to full-duty employment.

D. *Employee involvement*

1. Responsibilities of all employees

   - Report any work-related injury or illness to your supervisor immediately. (Failure to report a work-related injury or illness in a timely manner may result in denial of workers' compensation benefits.)
   - If released to return to work following a work-related injury with temporary, physician-imposed physical restrictions, report to work immediately and provide the company's workers' compnesation claims coordinator with the form on which the treating physician prescribed temporary physical restrictions. If seen by a physician during the normal business hours of this company, you must report back to work the day of the injury, unless otherwise instructed by your supervisor or the workers' compensation claims coordinator. If seen by a physician after the normal business hours of this company, you must report back to work the day after the injury (at your normal reporting time), unless otherwise instructed by your supervisor or the workers' compensation claims coordinator.
   - Adhere to the applicable physical restrictions imposed by the treating physician for the entire period for which those physical restrictions were imposed.
   - Continue to seek appropriate medical care throughout your recovery period, as directed by the treating physician and report any changes in temporary, physician-imposed physical restrictions (in writing) to the company's workers' compensation claims coordinator immediately.

2. Responsibilities of supervisory personnel
   - Document the physical requirements of the jobs of each subordinate employee and provide this documentation to the workers' compensation claims coordinator.

- Document specific tasks which may be performed by employees with common temporary physical restrictions, and provide this documentation to the workers' compensation claims coordinator.
- Assist the workers' compensation claims coordinator in identifying an appropriate modified duty assignment after an employee has been injured and released to return to work with temporary physical restrictions.
- Communicate the company's empathy and concern for each subordinate employee assigned to a modified duty position.
- Ensure that each subordinate employee assigned to a modified duty position knows his or her responsibilities to avoid any task that violates the physician's restrictions.
- Observe each subordinate employee assigned to a modified duty position to ensure that physical restrictions are not violated.
- Ensure that the workers' compensation claims coordinator is aware of the status, condition, and progress of each subordinate employee assigned to a modified duty position.

E.   *Program administration*

The following policies are have been enacted as elements of this Modified Duty Program to provide for the consistent application of the program and to prevent abuses.

1. Wages during modified duty assignments — Wages earned by employees engaged in modified duty tasks, assigned as a result of a work-related injury, will be based upon the work being performed, without regard to the employee's normal hourly or salaried wages, and will be consistent between all employees. In the event that a modified duty assignment is available only for a portion of the employee's normal weekly hours, it is possible that the employee will be compensated, through workers' compensation benefits, for the difference between the amount earned and the amount of workers' compensation benefits which would have been paid in the complete absence of a modified duty assignment. It is notable that workers' compensation benefits are based upon current workers' compensation legislation and are calculated as a percentage of the employee's average weekly income.

2. Refusal of Modified Duty Assignment — The company will make every reasonable effort to provide employees with modified duty assignments following a work-related injury for which the treating physician imposes temporary physical restrictions. Insofar as the modified duty assignment does not violate the treating physician's imposed physical restrictions, and insofar as the modified duty assignment is within the employee's physical and skill level capabilities, the employee is expected to return to work. Failure to return to work

without prior approval from your supervisor may result in termination of employment as it will be viewed as job abandonment.

F.   *Education and training*

To ensure that all employees are aware of the content of this Modified Duty Program and their related responsibilities, training will be conducted which includes explaining the objectives of the program, introducing the designated workers' compensation claims coordinator, and specifying each of the responsibilities of all employees, listed in Section D of this Modified Duty Program. This training will be conducted for all employees upon initial implementation of the program, for each new employee upon initial employment, and for all affected employees whenever there have been changes to the content of the program.

In addition to the above training content, each supervisor will receive training upon assignment to a supervisory position. This training will include an explanation of each of the responsibilities of all supervisory personnel, listed in "Section D" of this Modified Duty Program.

G.   *Program evaluation*

The workers' compensation claims coordinator is responsible for evaluating the effectiveness of this program on an annual basis. The effectiveness of this program will be based upon its ability to provide modified duty assignments for each employee injury for which the treating physician imposed temporary physical restrictions as opposed to time away from work. Further measures of this program's effectiveness is the demonstrable achievement of any or all of the enumerated objectives of the program, as listed in Section A of this Modified Duty Program.

Appendix E

National Resources for Alcohol and Drug Abuse Information

Workplace Helpline (800-WORKPLACE)
This organization provides technical assistance for employers and community organizations, including sample drug-free workplace policies, information on employee assistance programs, drug-testing information, and drug-free workplace resource materials.

National Clearinghouse for Alcohol and Drug Information (800-729-6686)
This organization offers more than 10,000 items on alcohol and other drug abuse (at no cost), drug-free workplace videotapes for both employers and employees, and the U.S. Department of Labor's Substance Abuse Information Database (SAID) on floppy diskette.

U.S. Department of Transportation (800-225-3784)
This organization offers regulatory information regarding DOT drug-free workplace transportation regulations and guidance documents.

U.S. Small Business Administration (202-401-3784)
This organization offers information to help small businesses implement a drug-free workplace program.

Employee Assistance Professionals Association (703-522-6272)

Employee Assistance Society of North America (810-545-3888)
These organizations are professional associations that represent employee assistance program providers. They offer information about employee assistance programs and referrals to available employee assistance program resources.

**National Association of State Alcohol
and Drug Abuse Directors, Inc.**   (202-783-6868)
This organization offers referrals to state government agencies responsible
for addressing alcohol and other drug issues.

Appendix F

Directory of State Workers' Compensation Administrations

Alabama

Workers' Compensation Division
Department of Industrial Relations
Industrial Relations Building
Montgomery, AL 36131
(334) 242-2868
www.dir.state.al.us/wc.htm

Alaska

Division of Workers' Compensation
Department of Labor
P.O. Box 25512
Juneau, AK 99802-5512
(907) 465-2790
www.state.ak.us/local/akpages/LABOR

Arizona

Industrial Commission of Arizona
800 West Washington Street
Phoenix, AZ 85007-2922
(602) 542-4411

Arkansas

Workers' Compensation Commission
Fourth and Spring Streets
P.O. Box 950
Little Rock, AR 72203-0950
(501) 682-3930
www.awcc.state.ar.us

California

Department of Industrial Relations
Division of Workers' Compensation
45 Fremont Street, Suite 3160
San Francisco, CA 94105
(415) 975-0700
www.dir.ca.gov

Connecticut

Workers' Compensation Commission
21 Oak Street
Hartford, CT 06106
(860) 493-1500
www.wcc.state.ct.us

Delaware

Department of Labor
Office of Workers' Compensation
Fox Valley
4425 North Market Street
Wilmington, DE 19802
(302) 761-8200

District of Columbia

Department of Employment Services
Office of Workers' Compensation
1200 Upshur Street, NW
Washington, DC 20011
(202) 576-6265

Florida

Division of Workers' Compensation
Department of Labor and Employment Security
301 Forrest Building
2728 Centerview Drive
Tallahassee, FL 32399-0680
(850) 488-2548
www.wc.les.state.fl.us/DWC

Georgia

State Board of Workers' Compensation
270 Peachtree Street, NW
One CNN Center
Atlanta, GA 30303-1299
(404) 656-3875

Hawaii

Disability Compensation Division
Department of Labor and Industrial Relations
P.O. Box 3769
Honolulu, HI 96812
(808) 586-9151

Idaho

Industrial Commission
317 Main Street
Boise, ID 83720
(208) 334-6000
www2.state.id.us/iic/index.htm

Illinois

Industrial Commission
100 West Randolph Street
Suite 8-200
Chicago, IL 60601
(312) 814-6555
www.state.il.us/agency/iic

Indiana

Workers' Compensation Board
402 West Washington Street
Room W-196
Indianapolis, IN 46204
(317) 232-3808
www.state.in.us/wkcomp/index.html

Iowa

Division of Industrial Services
Iowa Workforce Development
1000 East Grand Avenue
Des Moines, IA 50319
(515) 281-5934

Kansas

Division of Workers' Compensation
Department of Human Resources
800 SW Jackson Street, Suite 600
Topeka, KS 66612-1227
(785) 296-4000
www.hr.state.ks.us/wc/html/wc.htm

Kentucky

Department of Workers' Claims
Perimeter Park West
1270 Louisville Road, Building C
Frankfort, KY 40601
(502) 564-5550
www.state.ky.us/agencies/labor/wkrclaim.htm

Louisiana

Department of Labor
Office of Workers' Compensation Administration
P.O. Box 94040
Baton Rouge, LA 70804-9040
(504) 342-7555
www.ldol.state.la.us

Maine

Workers' Compensation Board
27 State House Station
Augusta, ME 04333
(207) 287-3751
www.state.me.us/wcb/wcbhome.htm

Maryland

Workers' Compensation Commission
6 North Liberty Street
Baltimore, MD 21201
(410) 767-0900

Massachusetts

Department of Industrial Accidents
600 Washington Street, 7th Floor
Boston, MA 02111
(617) 727-4900
www.state.ma.us/dia

Michigan

Bureau of Workers' Disability Compensation
Department of Consumer and Industry Services
P.O. Box 30016
Lansing, MI 48909
(517) 322-1296

Minnesota

Workers' Compensation Division
Department of Labor and Industry
443 Lafayette Road
St. Paul, MN 55155
(612) 297-4377
www.dolistate.mn.us

Mississippi

Workers' Compensation Commission
1428 Lakeland Drive
P.O. Box 5300
Jackson, MS 39296-5300
(601) 987-4200

Missouri

Division of Workers' Compensation
Department of Labor and Industrial Relations
3315 West Truman Boulevard
P.O. Box 58
Jefferson City, MO 65102
(314) 751-4231
www.dolir.state.mo.us/wc

Montana

State Fund Insurance Company
P.O. Box 4759
Helena, MT 59604-4759
(406) 444-7794

Nebraska

Workers' Compensation Court
State Capitol Building
P.O. Box 98908
Lincoln, NE 68509-8908
(402) 471-6468
www.nol.org/workcomp

Nevada

Industrial Relations Division
400 West King Street, 4th Floor
Carson City, NV 89710
(702) 687-3032

New Hampshire

Department of Labor
Division of Workers' Compensation
State Office Park South
95 Pleasant Court
Concord, NH 03301
(603) 271-3171

New Jersey

Department of Labor
Division of Workers' Compensation
P.O. Box 381
Trenton, NJ 08625-0381
(609) 292-2414
www.state.nj.us/labor/wc/Default.htm

New Mexico

Workers' Compensation Administration
2410 Centre Street, SE
P.O. Box 27198
Albuquerque, NM 87125-7198
(505) 841-6000
www.state.nm.us/wca

New York

Workers' Compensation Board
20 Park Street
Albany, NY 12207
(518) 474-6670
www.wcb.state.ny.us/col2.htm

North Carolina

Industrial Commission
Dobbs Building
430 North Salisbury Street
Raleigh, NC 27611
(919) 733-4820
www.comp.state.nc.us

North Dakota

Workers' Compensation Bureau
500 East Front Avenue
Bismarck, ND 58504-5685
(701) 328-3800

Ohio

Bureau of Workers' Compensation
30 West Spring Street
Columbus, OH 43266-0581
(614) 466-2950
www.ohiobwc.com

Oklahoma

Oklahoma Workers' Compensation Court
1915 North Stiles
Oklahoma City, OK 73105-4918
(405) 522-8600
www.state.ok.us/~okdol/workcomp

Oregon

Department of Consumer and Business Services
Workers' Compensation Division
350 Winter Street, N.E., Room 21
Salem, OR 97310-0220
(503) 947-7500
www.cbs.state.or.us

Pennsylvania

Bureau of Workers' Compensation
Department of Labor and Industry
1171 South Cameron Street, Room 103
Harrisburg, PA 17104-2501
(717) 783-5421
www.li.state.pa.ua

Rhode Island

Department of Labor and Training
Division of Injured Worker Services
610 Manton Avenue
P.O. Box 3500
Providence, RI 02909
(401) 457-1800

South Carolina

Workers' Compensation Commission
1612 Marion Street
P.O. Box 1715
Columbia, SC 29202
(803) 737-5700
www.state.sc.us/wcc

South Dakota

Division of Labor and Management
Department of Labor
Kneip Building, 3rd Floor
700 Governors Drive
Pierre, SD 57501-2277
(605) 773-3681
www.sate.sd.us/state/executive/dol/dlm/dlm-home.htm

Tennessee

Workers' Compensation Division
Department of Labor
710 James Robertson Parkway
Gateway Plaza, 2nd Floor
Nashville, TN 37243-0661
(615) 741-2395
www.state.tn.us/labor

Texas

Workers' Compensation Commission
Southfield Building
4000 South IH 35
Austin, TX 78704
(512) 804-4100
www.twcc.state.tx.us

Utah

Labor Commission
P.O. Box 146600
Salt Lake City, UT 84114-6600
(801) 530-6800
www.ind-com.state.ut.us/indacc.htm

Vermont

Department of Labor and Industry
National Life Building
Drawer 20
Montpelier, VT 05620-3401
(802) 828-2286
www.state.vt.us/labind

Virginia

Workers' Compensation Commission
1000 DMV Drive
P.O. Box 1794
Richmond, VA 23214
(804) 367-8633

Washington

Department of Labor and Industries
Headquarters Building
7273 Linderson Way, SW, 5th Floor
Olympia, WA 98504-4001
(360) 902-4203
www.wa.gov/lni

West Virginia

Bureau of Employment Programs
Workers' Compensation Division
4700 MacCorkle Avenue, SE
Executive Offices
Charleston, WV 25304
(304) 926-5048
www.state.wv.us/BEP/wc

Wisconsin

Workers' Compensation Division
Department of Workforce Development
201 East Washington Avenue, Room 161
P.O. Box 7901
Madison, WI 53707-7901
(608) 266-1340
www.dwd.state.wi.us/wc

Wyoming

Workers' Safety and Compensation Division
Department of Employment
122 West 25th Street, 2nd Floor
East Wing, Herschler Building
Cheyenne, WY 82002-0700
(307) 777-7159
www.wydoe.state.wy.us/wscd

Index

A

Abuse and fraud 7, *See also* Fraud
Accident investigations 65, 150–170
  accident prevention, understanding 167
  administrative and managerial deficiencies 168
  benefits 150
  compensability 153, 155
  documentation review 162
  documenting the scene 159
  environmental and physical conditions 159
  focus 155
  information gathering 157
  insufficient detail 195
  investigation report forms 158
  investigators 156, 170
  mitigating the causes 168
  near misses 155
  photographs 160
  preparedness 157
  revealing safety program weakness 152
  revealing trends 151
  root causes of accidents 166
  sketches 159
  surface causes 164
  trending 72
  unsafe acts 164
  videotape 160
  witness statements 161
  workers' compensation claims coordinator 115, 156
  workers' compensation premiums 155
Accident prevention, drug testing 177
Accidents
  causative factors 150
  investigating 16
  mitigating the causes 168
  rates 13
  records 31
  reports 163
  root causes 150
  trends 151
  witnesses 7
Actual primary loss 93
Administrative controls 17, 67, 168, 169
Administrative costs 9
Administrative deficiencies 168
Affiliation credit 99

B

Baseline worksite surveys 16
Benchmarking 70, 72
Bureau of Labor Statistics 31, 37
  Internet resource 80
Business image 23

C

CALOSHA 80
Carelessness 155
Carpal tunnel syndrome 215
Causative factors 150
CDC, *See* Centers for Disease Control and Prevention
Center for Substance Abuse Prevention 178
Centers for Disease Control and Prevention, Internet site 79
Change analysis 46
Charles N. Jeffress 18
Claimants
  fraud 189
  fraud red flags 197–200
Claims 103
  accident investigations 153, 154
  avoiding with safety management 12
  frequency 91
  informational 105, 114
  representatives 111
  responding to 210
  severity 91, 117
Classification codes 88
Closed claims 103
Code of Federal Regulations 77

Age of employees 36
Alcohol testing, compared to substance abuse programs 178
American Industrial Hygiene Association 83
American Red Cross 82, 83
American Society of Mechanical Engineers 3
American Society of Safety Engineers 83
American Standards Association 4
Americans with Disabilities Act 192, 205
  Internet site 80
Assigned risk pool 4
At-risk employees 36